ADVANCED ROCKER ARM TECH

Jim Miller

PHOTOS

Front Cover:	*Kuk Sool Won* student, *Pete Ferguson*, breaks a Coke bottle with an 18" high *Open Palm Strike* (1978) (Canon FTB, Ektachrome ASA-400 @ 1/1000 sec.)
Page 16:	Miller PRO-STUD™ (lower right) SB Billet Rocker Arm
Pages 23/96:	Miller MID-LIFT Ford NASCAR Pro-Shaft system (2000)
Page 128:	PVS Extruded bar of 7075-T6 Aluminum (upper left) PVS PA4570 Extruded Rocker (middle right) (2003)
Page 162:	Miller PRO-STAND™ (lower left) SB Billet Rocker System
Page 167:	BOSS 429 Series III Miller MID-LIFT™ Rockers (1983)
Back Cover:	All aluminum 9,300 RPM *BOSS 429* (MRE, 1983)

Photographs by ***Jim Miller***

1603151953

RETAIL PRICE: $87.50 USD

ADVANCED ROCKER ARM TECH
Jim Miller

CONTENTS

PREFACE

Jim Miller developed *precision rocker arm geometry* as a teenager in 1973 and Patented it in 1982, which became known as "MID-LIFT." As an engine builder not interested in the chores of in-house manufacturing, it would be more than a decade before Jim founded Miller Engineering Inc (MEI), and the Patent would finally be used to develop custom, precision rocker arm systems for more than an elite few.

By 2000, MID-LIFT™ had been used across several rocker system platforms, from stud mounted aluminum and steel designs, to stand mounted systems in all forms of amateur and professional racing from around the world, including NASCAR. Along the way Jim would earn three (3) more valve-train Patents, plus pioneer novel manufacturing.

In 2003, the MID-LIFT.com web site was launched to provide technical information to engine builders on the "basics" of *installed* rocker arm geometry for any brand they used, while also introducing the difference of MID-LIFT systems. In 2006, for the more professional needs there was an "advanced tech" section made available by paid subscription that not only went into far more detail by including the *design* geometry, but also the rocker arm's integration with *camshaft* and *induction system* decisions.

For over 20 years the competing cam and rocker arm companies denounced MID-LIFT™ geometry, and engine builders were caught between fact and profits that established companies didn't want to lose denouncing their old standards. Today, most have adopted Miller's geometry, even though they don't admit it, and some even try to take credit, referring to "Half Lift."

In 2010, Jim published the *Advanced Tech Bundle*; a compilation of several tech articles specific to the topic each reader required and sold individually as PDF downloads. These covered: *Installed Geometry, Design Geometry, Cam Basics,* and *Valve Train*; plus *Terms & Definitions*, with over 225 items that often included expanded tech.

Advanced Rocker Arm Tech, is a completely revised and expanded version of the *Advanced Tech Bundle*; rewritten by the founder of MID-LIFT Geometry™ in a casual, no nonsense voice with precise illustrations, in-depth technical commentary and historical engine systems analysis. Now retired from manufacturing, the author candidly reveals many confidential insights acquired from over four decades of pioneering innovation and experience as a leading valve-train designer and cylinder head developer who quietly consulted and worked for some of the biggest names in professional racing.

This is a <u>*must read*</u> publication for both professionals and novices alike, who are serious about extracting the most from their race engines and/or parts development programs. *This publication is for educational purposes only, and not intended to endorse, or induce sales of any specific brands.* We hope you find this book an indispensible resource for optimizing your investment in high performance engines, racing, or product development.

—Editor.

AUTHOR'S FORWARD

Everything in the valve-train, except the revolving cam, reciprocates in a near [linear] path. The rocker arm reciprocates too, however its fundamental dynamics follow a radial path that rotates about a fixed axis. As simple as this distinction is, it is critical to the proper design and installation of the rocker arm to function efficiently with its integral linear motion components.

The reciprocating valve-train components are dynamic; spanning a wide range of *action and reaction* against forces often overlooked or underestimated; too often "fixed" by adding more valve spring pressure, only to mask one symptom while creating another. Aggressive cam ramps running at high RPMs with valve springs generating the second highest temperature element of the engine create and magnify HARMONICS which escalate inherent frequencies already generated by tappets, push-rods, rocker arms and valve to seat dancing.

What is harmonics? It has been described many ways; some inaccurately. Here's my definition:

Harmonics in a valve-train is the oscillating waves created by objects connected to one another which develop individual frequencies that often collide at various high RPM points. They coexist innocuously until a specific threshold is passed; then the cam follower, pushrod and rocker arm each reach a frequency at higher RPMs where their separate harmonics become dominated by the worst offender: the valve spring. Valve springs are a *twisting torsion rod* that just happens to be spun in a circle. Once valve spring harmonics pass a controlled threshold the chaos on the entire valve train is not only innate, but perpetual. Each component influences valve spring harmonics; but none more than <u>rocker arms</u> with *bad geometry, excessive loads and motion*.

Valve-train geometry is at the heart of controlling harmonics and optimizing the valve-train. Either the camshaft exclusively controls your valve timing with minimal wasted motion and energy of the linkage in between; or the camshaft's design is compounded by a rocker arm operating with wasted motion that robs energy and increases the entire valve-train's harmonics.

Mid-Lift geometry standardizes the critical elements of motion so an emphasis on accurately quantifying and developing the cam timing to optimize the head's maximum volumetric efficiency can be accomplished without futilely chasing erroneous symptoms.

This book sells no products; makes no promises; states no final solutions as the "be all" end to every engine building problem. Even though I have many terms explained for the first time engine builder or novice, I am assuming the reader is familiar with most of the parts and uses mentioned, so I have refrained from filling it with a lot of product photos.

This book should arm you with a point-of-view earned over many years from many mistakes of what does not work, as much as what does. When you know what's important, Mid-Lift geometry can be easily engineered into the valve-train of any OHV engine. At the very least you should have a better understanding of what your existing valve-train is doing, and why; and how to correct it with the least amount of costs.

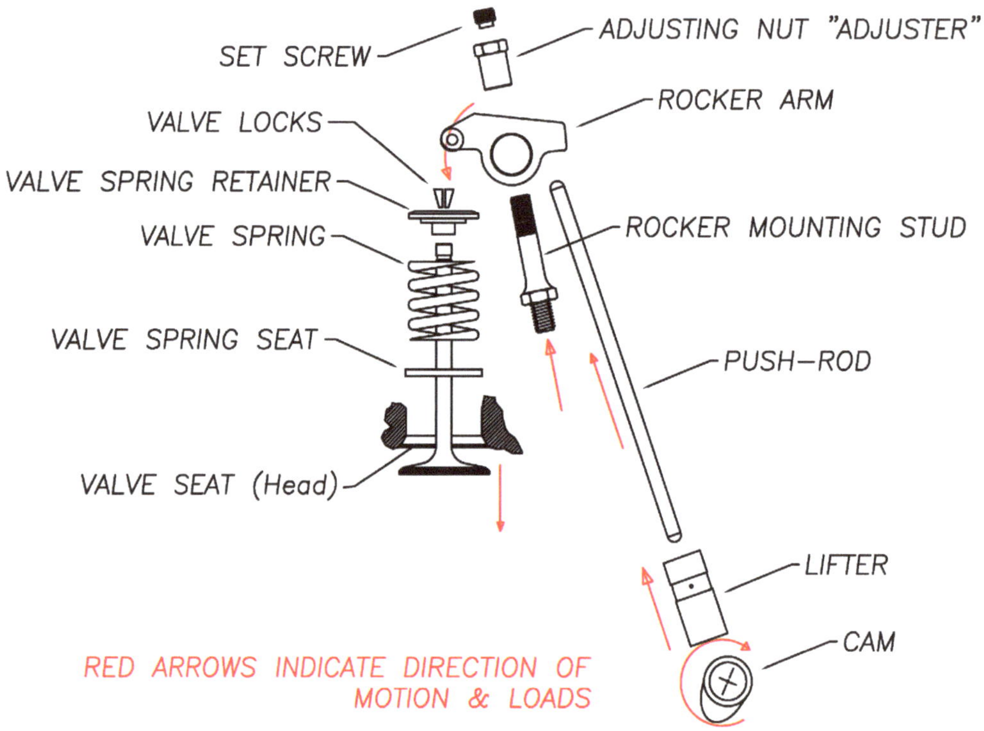

VALVE-TRAIN
FUNDAMENTALS

BASICS 101: All Over-Head Valve (OHV) engines have pushrods that link the **cam** and **tappet** information to the **rocker arm**. The rocker arm pivots about an axis on a shaft mounted to the cylinder head, pushed on one end by a pushrod, with the rocker's opposite end pushing down upon a spring loaded valve tip, that reciprocates. I know, this is "engines 101" – but it's worth placing in perspective.

I emphasize in several chapters that there are two additional perspectives that need to be understood: 1. "Design Geometry" and 2. "Installed Geometry." These two distinctions refer to the operating characteristics of each end of the rocker arm, which are not symmetrical. Each end has an effect on the efficiency or failure of all components in the valve-train. Except for the cam, everything in the valve-train is reciprocating; back and forth; sixty, seventy times a second at high RPMs. The "cam" is a rotating device for which its very term announces an eccentric lobe, where anything sitting atop it will rise and fall to follow its profile. This is a gross over-simplification of an operating process that is very, *very* precise and exact. The running surface of the cam lobe is ground to a precision that's measured in ten-thousandths of an inch (0.0001"), but the program that determines the shape of this surface is calculated in millionths to stacked dimensions only a computer can resolve as this *lobe* changes the speed, time, and precise movement of whatever is riding atop this surface. Especially if it is a "wheel."

The component that rides upon the cam is known as a "tappet" or "lifter" or "**cam follower**," depending on who's talking. Each term has its own etymology in the American language of engine talk, but the part and its task is the same. Riding atop the cam follower is the **pushrod**. The last component linking the cam to the valve is the rocker arm; and when considered with the valve, we now have our "valve-train."

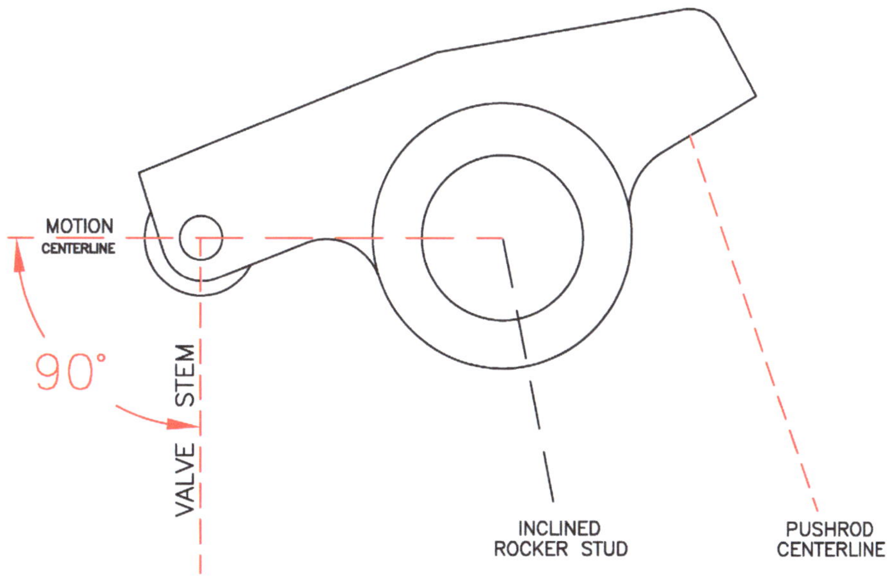

CONVENTIONAL ROLLER ROCKER GEOMETRY

ROCKER GEOMETRY being set accurately is deadly critical to the valve-train's efficiency. The above is ***NOT*** how the conventional roller tip rockers have been installed, nor designed. This 90° relationship would have been reached at the 3/4 lift, or worse... full lift. But this is how it **should be.**

What is Rocker Geometry? Rocker geometry is the angular relationship of its pivot points with *each other* as the rocker reciprocates about its axis with respect to the driving and driven linear operating components it connects with, on an Over Head Valve (OHV) engine (non-overhead cam). Simply, all pushrods lean into the valve at a specific angle. Some engines (i.e., Chrysler Hemi and BOSS 429) the pushrods lean *away* from the valve. Either way, the rocker arm reciprocates between these two components and converts the "linear" motion (tappet and pushrod) into "radial" motion; multiplies this by a "ratio," then translates its radial motion back into linear motion pushing upon the valve.

Because the rocker is a "radial instrument" an arc is created where the pushrod and valve connect. Depending on the rocker arm's design geometry, which determines the pushrod's path, this arc will influence how much of the CAM'S linear information is translated. Predominantly, the cam "rate" (velocity) and "duration" (time) is affected by ***where*** that rate and duration occurs! In other words, the rocker arm's angles either magnify or dilute what the cam tells it, at different points of cam lift; and this applies to all THREE parameters the cam is defined by: LIFT, VELOCITY and DURATION.

The above illustration shows a conventional rocker arm installed properly using the VALVE SIDE to yield a 90° angle to the valve between the roller to trunnion axis`. Most of the largest stud mount rocker manufacturers, as well as the most well known shaft rocker arm manufacturers promote a geometry that is more like the 1/3 rule (shown on p.9), and subsequent pages for a "shoe tip" rocker, even though their designs are a roller tip. This is WRONG. Make no mistake, anything other than setting VALVE side as shown above will waste motion on the valve, add additional side loads to the valve guide, and worse: amplify "harmonics" – which isn't necessary – and destroys both RPMs, valve springs and worse.

NOTE: See "Installed Geometry" (p.16), and **RADIAL ARCS** (p.12) for dimensions for your application.

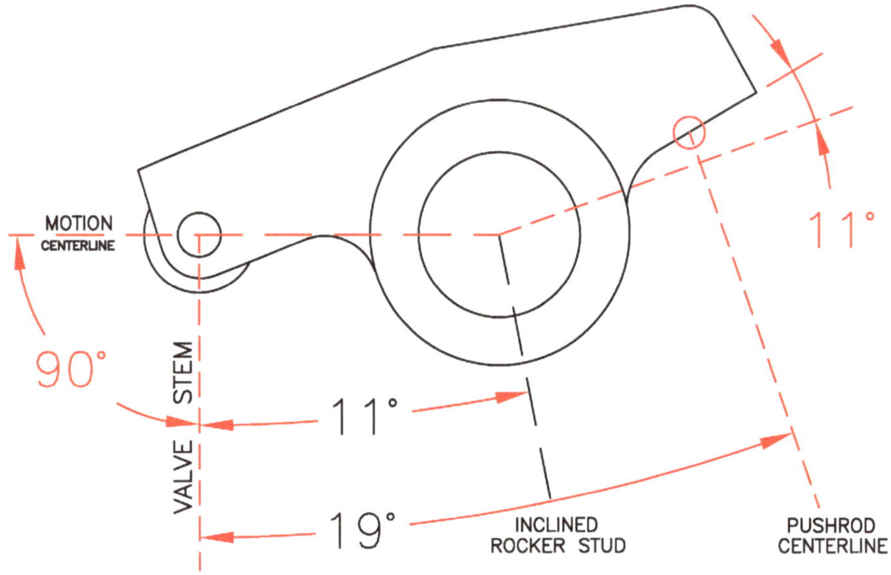

CONVENTIONAL ROLLER ROCKER GEOMETRY

This rocker silhouette, same as that shown on the preceding p.7, now depicts the problem with conventional rocker arms where the pushrod cup as TOO High when the roller tip end has been set to MID-LIFT. The pushrod has already passed through its mid-lift point and is now 11° beyond.

Considering the rocker arm only rotates approximately 26° total (SB Chevy at .600" Valve Lift), and only HALF of this total 26° (13°) is used to set mid-lift, then being 11° PAST mid-lift is nearly double the error and definitely a waste of motion and energy. This amounts to approximately an 85% error. This 11° error was not pulled out of space or any other dark place. It was measured from samples of some of America's top selling stud mount rocker arm brands. As unwanted as this is, and even though the pushrod is over-arcing, it is still **THE LESSER OF EVILS** to use for your system.

As noted on p.30, this pushrod error has a "wrap-around" effect, where most of the cam information will be conveyed to the rocker, in spite of such an over-arc. But the harmonics that is created is lethal.

If the rocker in the above illustration was rotated clockwise 11° so that the bottom of the pushrod cup lined up with the 90° pushrod tangent at mid-lift, you would then see the BOTTOM of the Roller now moved up to sit on the 90° "Motion Centerline," instead of being at its CENTER, where it is shown now. That's a big error. The effect this change would have, would be to improve the cam information reaching the rocker, *but...* at the expense of an over-arcing roller tip, which would now be imposing increased side loads on the valve stem and guide, and wear atop the valve tip; and *reduced* acceleration of valve.

The reason for how this "half roller" geometry mistake was made, then copied by the entire rocker arm industry is explained in the ***History Of MID-LFT™*** (p.163). The chapter on DESIGN GEOMETRY, you can see what needs to be measured and changed to fix this.

~jM

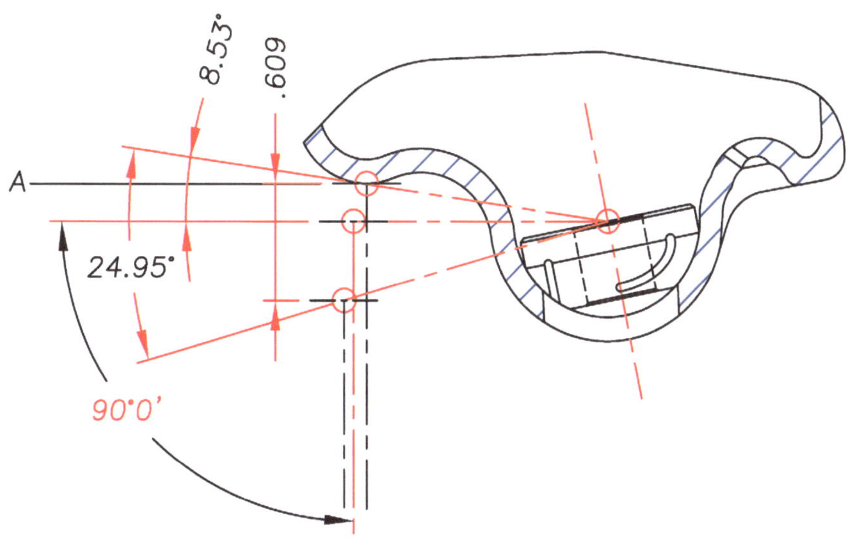

THE ONE-THIRD (1/3) RULE

The 1/3 Rule, as it has always been known, is shown above. This pertains to a "shoe" tip rocker arm; not a roller tip rocker. The principle of this geometry measures from the actual contact point of the rocker's pad with the valve tip (the little circles identify). This rule establishes that at 1/3 of the total valve lift, a perpendicular (90° angle) is made between the valve and the rocker arm's fulcrum.

The **small RED** circles show where the contact points (datum) will be at **ZERO LIFT**, **MID-LIFT** (90° to the fulcrum at 1/3 Peak Lift), and the lower datum is **FULL LIFT**. Notice the increasing length away from the fulcrum of these last two datum points. All else being equal, the rocker ratio is INCREASING and valve speed is accelerating. Not Good! Page 11 shows the full "scraping" across the valve to be .121".

The reasoning for the 1/3 rule has never been accurately described in layman's terms, but it has to do with averaging out the shifting change in ratio. On the following pages I've put together a series of illustrations that show this in all three stages, showing the radial circles created by these three contact points. These radial circles are the path that the rocker arm rotates at EACH of these three lengths, BUT... as you will see from the red lines drawn between the tangent circles, this dimension is constantly increasing from closed valve to full lift, as set in this geometry.

There is no physical correlation between a "shoe tip" rocker arm and a "roller tip" rocker arm when it comes to cause and effect. Yet, it is the shoe tip rocker arm that was the impetus for the racing industry's gross mistake on accurate geometry when it first began making roller tip rocker arms half a century ago. Stubbornly, so much money was made by the top cam and rocker arm companies, that when they learned of a better way (MID-LIFT), they stuck their heads in the sand and kept delivering over-arcing rockers to unsuspecting engine builders. A practice, that unfortunately still continues.

That mistake is explained in the ***History Of MID-LFT™*** (p.163). The following two pages depict the stages of lift using this 1/3 rule, to illustrate the intent and effect.

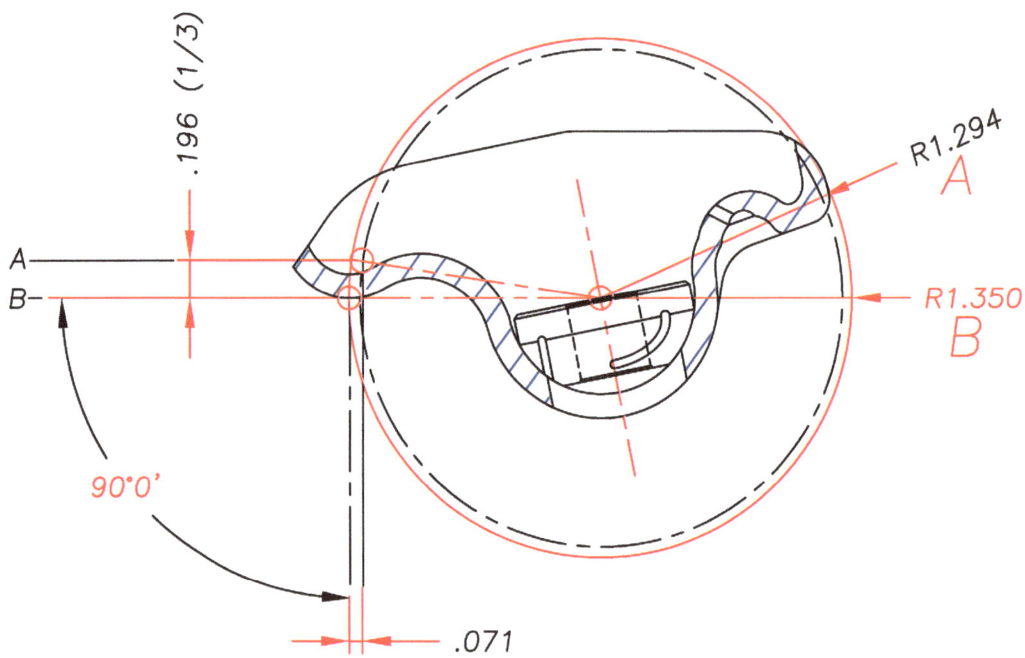

THE ONE-THIRD (1/3) RULE

TRADITIONAL 1/3 RULE is shown here with the second radial.

The **LARGE BLACK** CIRCLE (**A**) represents the active contact LENGTH in the CLOSED valve position, while the **LARGE RED** CIRCLE (**B**) represents the DESIGN length of **1.350"**. The rocker is shown at 1/3 lift. A **.070"** wear pattern will be scuffed across the valve tip for this mere .196" valve lift (upper left).

The ever critical design length radial of the small block Chevrolet rocker is shown in the 90° relationship to the valve, as represented by the **BLACK** vertical centerline. The above illustration depicts the axis of the ball fulcrum at the top of the ball fulcrum, which the PVS rockers utilized, because their fulcrums were custom made and not interchangeable with an OEM rocker. It made setting geometry easy, even though the PVS rockers already had a top flat to reference setting installed geometry from. The above rocker silhouette is OEM.

Setting geometry on an OEM shoe rocker arm is difficult, because the ball fulcrum of the OEM style rockers is NOT half its height, or equal to half its diameter. The axis is some odd value above the OEM ball fulcrum's top (like .050" or .060", depending on brand). Finding the exact axis of the BALL on an OEM style *shoe tip* rocker arm would require making a washer-spacer to add to this short dimension. Then mounting to the stud and elevating it with an adjustable pushrod while measuring to the tip of the valve would be much easier to set by. If you build many class type engines that require a "stock type" rocker, then a jig which had this additional height built into would be best for quickly establishing where the adjusting nut should stop and how long the pushrod should be, for any given valve lift. But you would NOT set up for the 1/3 rule. You would naturally set up for MID-LIFT. The next illustration (p.11) depicts the final step.

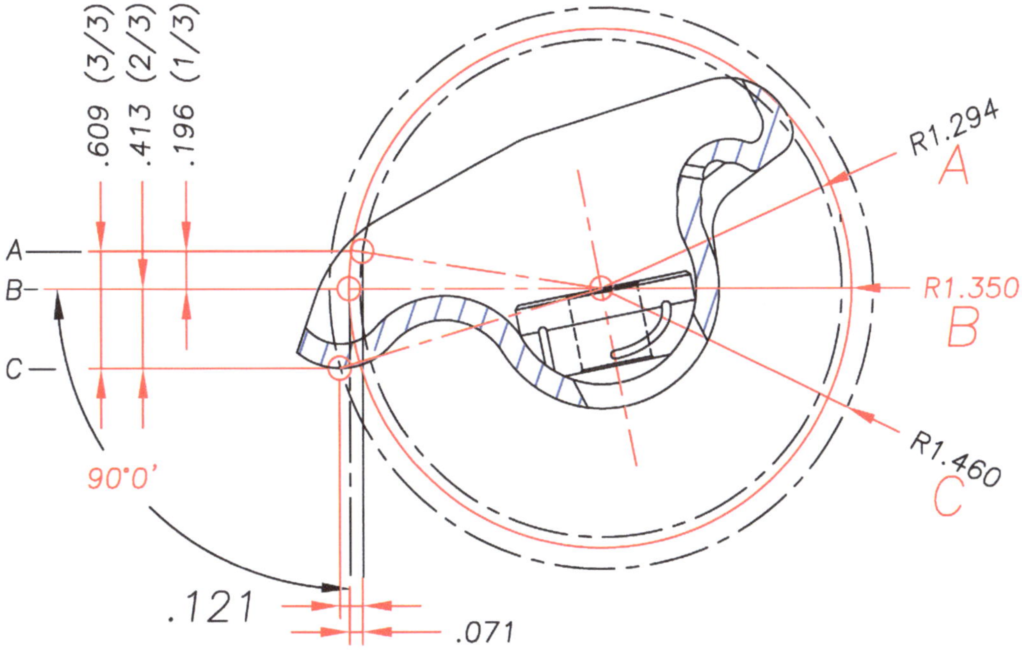

THE ONE-THIRD (1/3) RULE

The **LARGE BLACK SOLID** CIRCLE is (**A**), the shortest radial with a length, at 1.294" at .196" VL.
The **LARGE RED** CIRCLE is (**B**), and is the DESIGN RADIAL of 1.350" at .400" VL.
The **LARGE BLACK DASHED** CIRCLE is (**C**), the longest radial with a length of 1.460" at .608" VL.

This is a "**radial stretch**" of **.166"** (NOT good). Notice how this translates to **.121"** VALVE TIP SWEEP, as measured perpendicular to the VALVE (shown in the lower left). The effect of this is a direct side load bias *away* from the rocker axis that overloads the rocker stud as velocity increases toward full lift.

Depending on how the pushrod geometry is established with the rocker arm, a 1.50:1 rocker ratio, without a correction factor built in, would have a pushrod cup distance from the fulcrum of .900". If you determine the ratio based on this, then LENGTH "A" would generate a ratio of 1.43:1, LENGTH "B" would be our design ratio of 1.50:1, and LENGTH "C" would be 1.62:1. Lower the rocker reduces this.

Measuring rocker arm geometry under this rule is not logical. In fact, the entire principle of measuring a shoe tip rocker arm by its contact surface, is also something I never agreed with, and the success of my PVS OE Series rockers proved this. But it is the *only* way to set existing OEM designs, but not at 1/3 lift.

> **NOTE:** The roller tip rocker arm follows this arc around an axis, but because of the "floating" contact point of the roller's outside diameter upon the valve tip its control and limit of this stretching deviation is much less. Even when installed arbitrarily, as most roller tip rocker arms are, this phenomena between the two concepts makes the valve lift at a different rate throughout its lift cycle, and THIS is the real culprit to the majority of differences in a test between the two designs, not "needle bearings," not "aluminum," not "name brand," or any of the usual hype used to sell roller tip rocker arms.

~jM

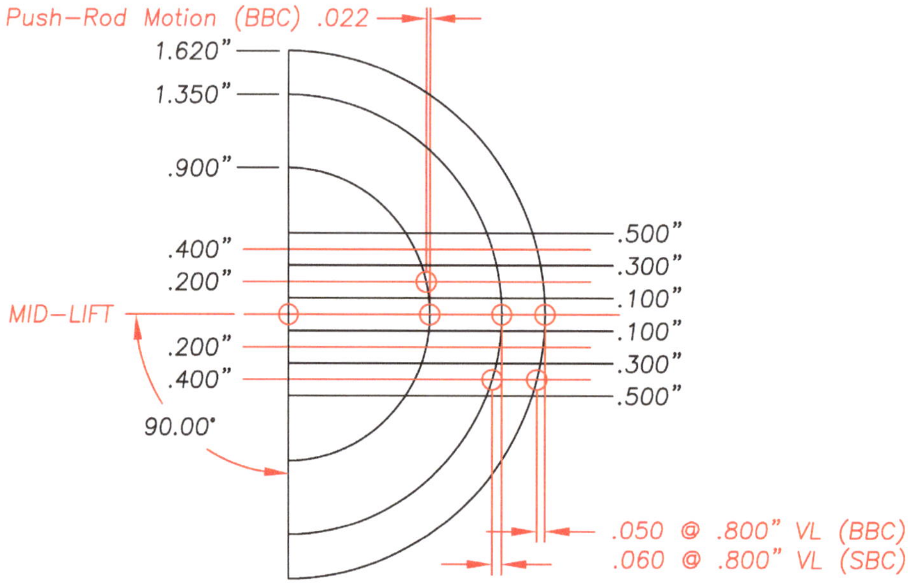

RADIAL ARCS

RADIAL ARCS are illustrated throughout this publication for a fundamental point of view for what is important to focus on with rocker arm design and installation geometry. Radial arcs are simply the paths of the rocker's pushrod tip ***axis*** and roller tip ***axis*** as each reciprocate about the rocker arm's shaft axis.

The above illustrates three (3) radial dimensions on one graph, so each should be viewed without confusion except for comparing the effect of their lengths, which is given by the **RED** text values.

Two radial arcs represent rocker lengths (roller to trunnion) of 1.620" for the BB Chevy and 1.350" for the SB Chevy. Each are measured at .800" Valve Lift, marked by the **RED** tangent circles.

The third radial arc of .900" represents a pushrod cup to trunnion length, marked for a cam lift of .400". Please note that the tangent circles are only grabbing *half* of these values; .200" for the cam lift, and .400" for the valve lift. The principle of Mid-Lift is always measured from the middle point of motion, as the companion tangent circles to each measurement show by their attachment to the MID-LIFT line. So only *half* the value is used for whatever valve lift or cam lift is being measured. (See rocker view, p.33.)

MID-LIFT is always measured as only HALF the total value, since the second half of cam and valve lift motion will duplicate the first half if the arcs are divided equally. The vertical line in this illustration should be considered the centerline for the *valve* on any rocker length arc; or the *pushrod* centerline for any pushrod arc. These must **always** be at 90° to whichever you are measuring; valve or pushrod.

As you can see the CAD measurement for the BBC roller tip motion on the valve is only **.050**" for .800" valve lift. The SBC is only .060" for .800" VL (which is more VL than what would normally be usable). The Pushrod arc is only .022" for the cam/tappet lift of .400". If you know your rocker arm's dimensions, you can duplicate this with any accurate drawing program, and predict the exact minimum motion you can achieve for either the roller tip or the pushrod. **See: *INSTALLED GEOMETRY*** (p.16).

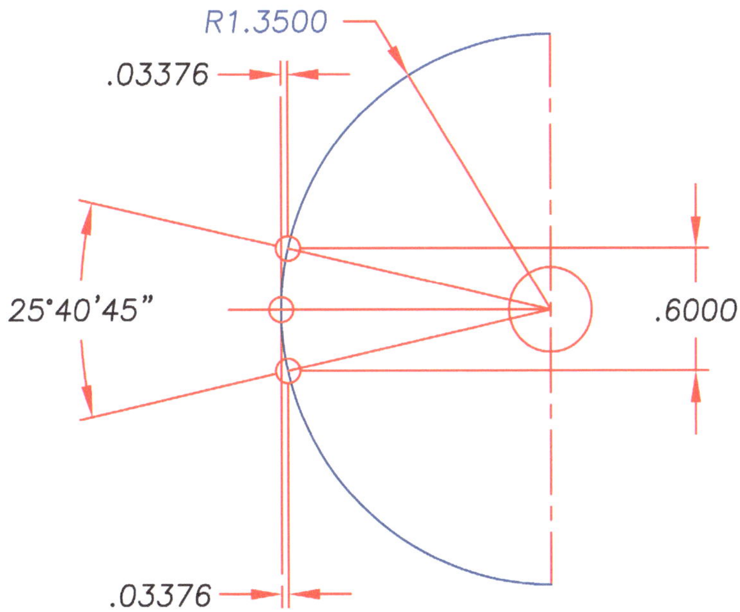

RADIAL ARCS

RADIAL ARC of 1.350" Trunnion to Roller pin axis length is shown above, and is *only* the roller tip side of the rocker arm dimension. This is all that's important for setting INSTALLED Geometry. Setting the rocker at mid-lift to divide the arc equally is the only way to achieve such a minimal in-and-out motion as this .033" shown here for .600"VL. This is all that's important for setting INSTALLED Geometry.

This angle of rotation is only 25°40'45" (25.67° is the decimal value). Most of the angular illustrations in this book will be in decimal, but here I have used the engineering standard of degrees/minutes/seconds, and let the decimal values run out to the fifth position. Such accuracy (easy for CAD) is not needed in the real world, but in the CAM world this is standard operating procedure. Remember that the rocker arm is downstream of all the cam's precision instructions. So being slopping about pushrod length, rocker arm pivot height to the valve, and fundamental rocker arm design criteria is all critical to getting the cam's accuracy conveyed, and multiplied through to the valve.

If you miss mounting the rocker arm's height to the valve by only .050", which is an easy thing to do, then that error is flipped on the other side of MID-LIFT. I'll say it over and over, your whole mission in setting geometry is to divide the arc evenly, and you only have to measure the first half to do this. But missing it on the first half of valve lift, will move that error (whether it is too low or too high) to the other side of the valve lift. This changes the degrees of cam information and where they occur in the lift cycle.

Such an error was part of a Dyno session to test cams, obviously a potentially good cam would be getting a bad rap. In the other extreme, an overly aggressive cam would be tamed down by this excessive motion that did NOT have the above, and a bad cam would be seen as acceptable. This is why establishing the rocker arm's installed geometry for MID-LIFT is crucial from test to test, head to head, and cam to cam if you really want to know what is going on with your engine.
The next chapter covers this.

~jM

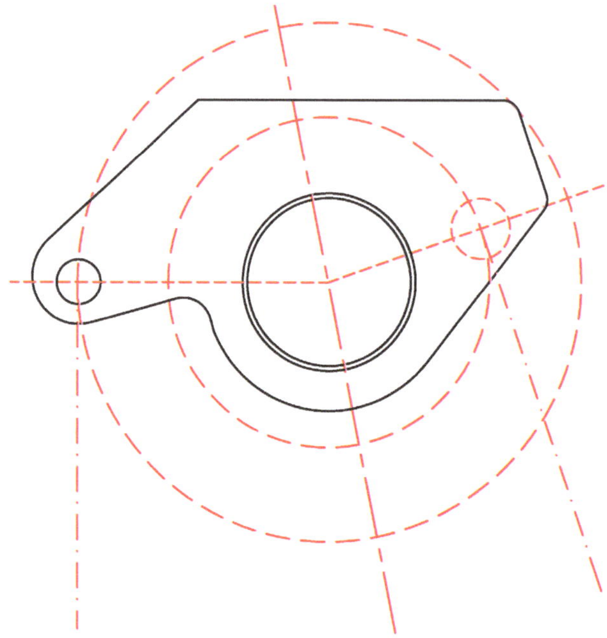

MID-LIFT Geometry™

WHAT IS MID-LIFT?

DESIGN and **INSTALLATION**. These two simple perspectives of rocker arm function must be understood separately. Because you can have one and not the other. If the "design geometry" isn't correct, then your "installed geometry" can only be set for the lesser of two evils: either setting the pivot points for a **MID-LIFT** relationship on the "valve" side of the rocker arm, or the "push rod" side of the rocker arm; but NOT both. On a stud mounted system, this is done by changing the pushrod length. On a stand (shaft) system, this is done by adjusting the stand's height. Accordingly, these two perspectives dictate TWO kinds of geometry:

The first is: "***Design Geometry™***."

The second is: "***Installed Geometry™***."

This is not rocket science. To the contrary, it is actually quite simple. But understanding the difference between **DESIGN** and **INSTALLED** geometry will simplify what you need to do for not only choosing the right parts, but also installing them correctly. It will save you a lot of grief, time and money in the process. So let's go back to rocker arm basics; what MID-LIFT is, and where it came from...

Note: This chapter, as with several chapters of *Rocker Arm Advanced Tech,* has been revised, updated and expanded from the original compilation of individual articles originally published in *The Advanced Tech Bundle*. Accordingly, this chapter may contain some repetition of points from other chapters. For continuity most of these have been edited wherever possible, but some have been left intact for emphasis where relevant.

WHAT IS MID-LIFT?

What really matters with the driving force of the rocker arm, is the PUSH-ROD. All OHV engines have an inherent angle between the pushrod and the valve, which dictates many factors of the rocker arm's motion. On some engines, this angle leans away (called a negative attack angle), but on most American made engines, this angle leans IN toward the valve centerline (positive). So in keeping with a simple concept of minimum "in-and-out" motion for the pushrod, we place the axis for the top of the pushrod at a 90° angle, to the axis of the rocker arm itself. But to make the rocker arm truly MID-LIFT in "Design Geometry," it has to do this same thing on the opposite side, where the VALVE is.

If either side of the rocker arm does not do this, then one side or both (depending on the "Installed Geometry"), will move out of its radial path more than it needs, and the consequences will be wasted "linear" motion from the cam to the valve; excessive harmonics at high speed; extra contact heat from extra friction within the socket, and finally extra side loads being applied in the direction where the linear deviation occurs. In example, if the pushrod is too high in the rocker (a common trait of most designs that still exists), then as the cam lifts the tappet and pushrod up in the block, the rocker end tip of the pushrod which is "floating" with the rocker arm's motion will push upward and in toward the stud. As it leaves this straight up path and begins going around, then in toward the stud, it will "slow down" the rocker's motion as it becomes most acute, because it is not following the cam and tappet in line where maximum velocity occurs. However, if this deviation is not too excessive, then the "wrap-around" effect minimizes cam loss; but *harmonics* is much higher to this slapping around motion. This is ONLY the pushrod side. The valve tip side has its own chaos if it is not set to MID-LIFT with the valve. The height of the roller tip ABOVE the TRUNNION (or shaft, for stand mount rockers) at zero lift determines installed geometry, and it must be half of the NET VL. If it over-arcs from being too low, then the best it can do is rotate the same amount of degrees that the cam side has dictated, and net lift will be less. (See: Over-arcing illustration on p.20, then compare with MID-LIFT illustration on p.22).

MID-LIFT Geometry™

As explained in the ***History Of MID-LIFT™*** (p.163), the concept of MID-LIFT geometry began in 1973. By 1974, it was finalized into the principles of precision geometry that was awarded a Patent in 1982; the only such Patent ever issued on a "concept" for rocker geometry. The Miller MID-LIFT™ Patent has since been cited many times in Patent applications from around the world, because it set a precedent.

Over the years through social media and blogs, I've seen occasional references erroneously conclude *mid-lift* concepts were used on engines long ago; from airplanes to foreign cars. That's a myth. *Never before* was the concept of equally dividing the axis tangent points on *both* sides of the rocker arm at *mid-lift* ever incorporated into OHV engines, *anywhere.* If they had, the **MID-LIFT *Patent*** wouldn't exist.

These Patent winning concepts have never wavered or changed, and they established not only the "standard" by which rocker arm geometry is defined, but also the standards of other valve-train components that revolve around it; such as valve tip heights; valve lengths; stud placement and other necessary dimensions for proper integration of the valve-train with cylinder head preparation; a mandatory consideration for anyone serious about avoiding conflicts with other valve-train components.

The following is a fundamental (but necessary) overview of the following components and their integration with companion parts. ***CAMS, INSTALLED GEOMETRY*** *and* ***DESIGN GEOMETRY*** are each covered with expanded tech in subsequent chapters, including ***TERMS & DEFINITIONS***.

~jM

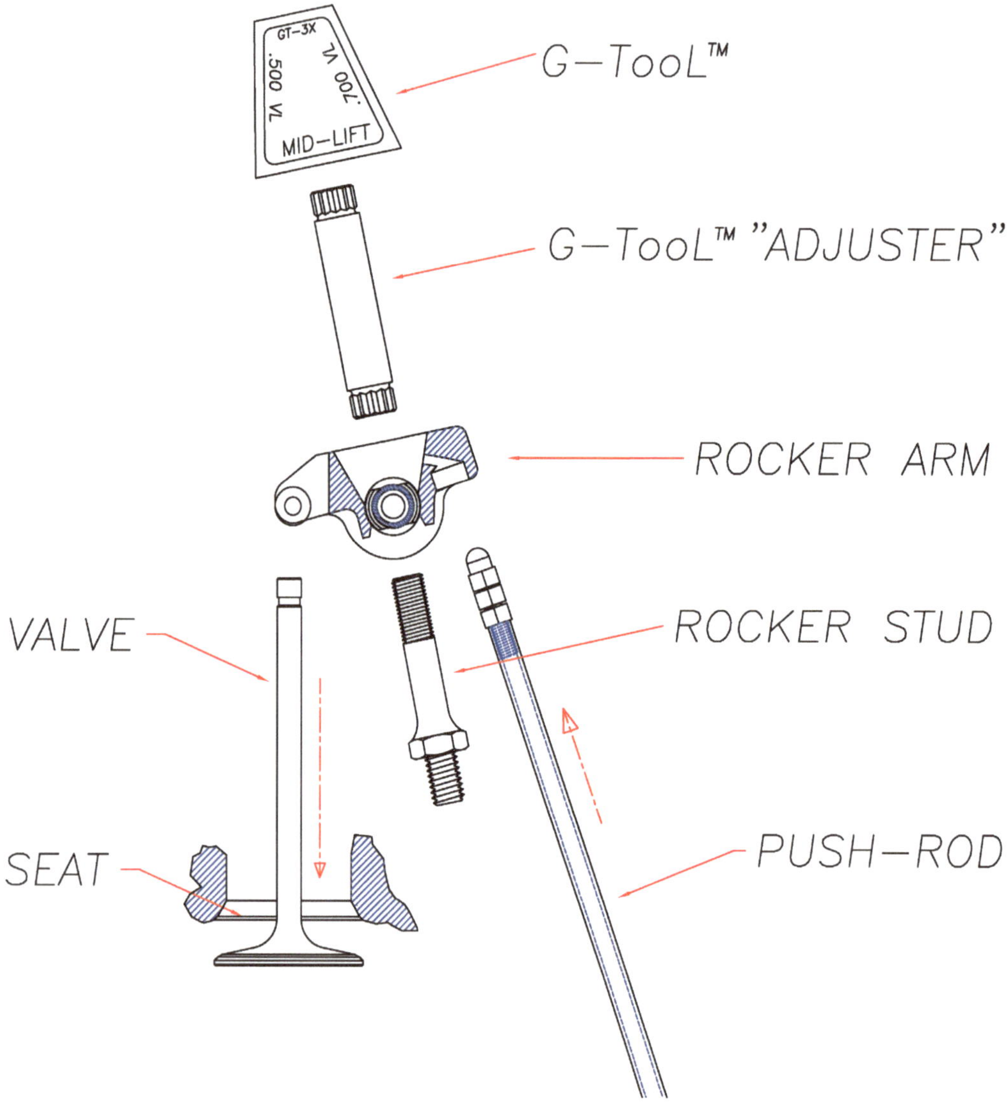

INSTALLED GEOMETRY™

STUD MOUNT ROCKER ARMS

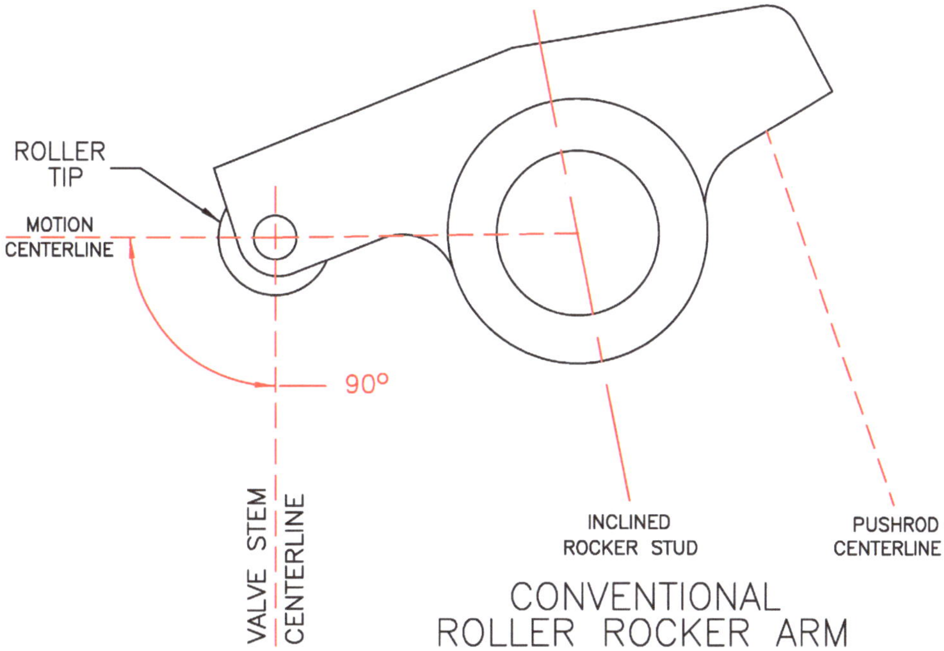

INSTALLED GEOMETRY™
STUD MOUNT ROCKERS

INTRODUCTION

FIRST PUBLISHED in the 1980s (with less detail)... this is the definitive instruction for how and why to set your roller tip stud mount rocker arms up for the least amount of wasted motion on the valve, the least amount of side loads, the least amount of wasted *radial* motion from the rocker's valve tip end, and the least amount of costs; rather than confusing gimmicks, tools and other methods that fail to be precise and waste money.

1. **YOU NEED A SHORT STRAIGHT EDGED RULER (or equivalent).**

2. **YOU NEED TO KNOW YOUR NET VALVE LIFT.**

3. **YOU NEED DIAL CALIPERS (or the equivalent).**

If your rockers are the correct length for your heads, as measured from the stud to the valve, then the following is all you need to know for setting your INSTALLED Geometry.

Note: This chapter may contain some repetition of points from other chapters. For continuity most of these have been edited wherever possible, but some have been left intact for emphasis where relevant.

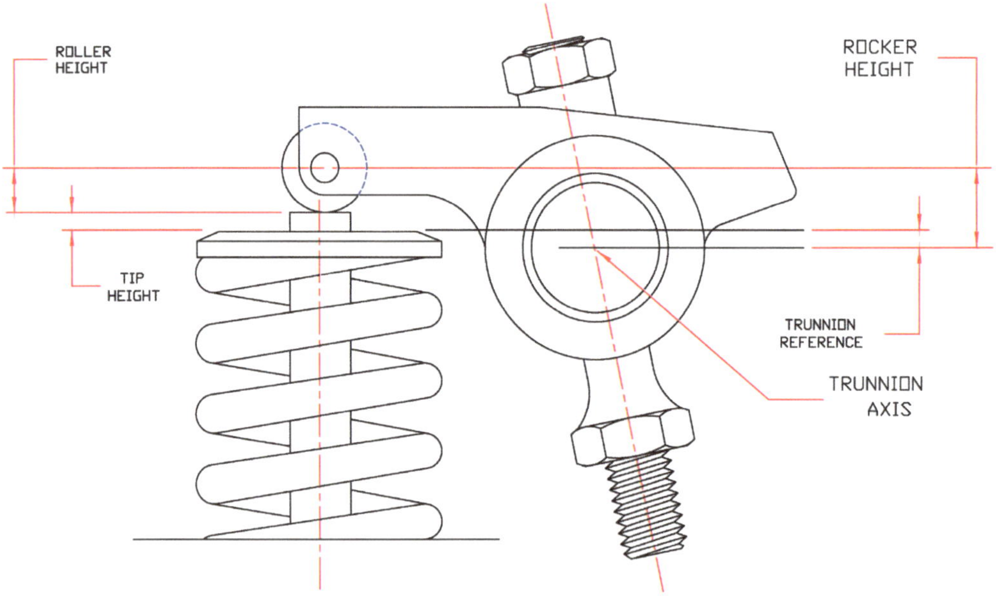

SETTING INSTALLED GEOMETRY

INSTALLED Geometry™, is the art of adjusting the rocker's axis below the roller tip centerline, exactly HALF of whatever the NET valve lift is. Here is the illustration that Jim Miller published in magazines several years before he used it in the 1988 Lunati catalog, where it remained for 12 years. With the valve closed, the two key measurements to the upper LEFT take reference from the valve spring retainer and together create a "***STACK HEIGHT***" which is then measured in comparison to the trunnion's centerline to establish the "**ROCKER HEIGHT**". *When set properly, this height will be HALF of your NET Valve Lift, as measured in the CLOSED VALVE position.*

By using the VALVE SPRING RETAINER as a reference plane, you can run a machinist ruler (or similarly accurate tool) along the side of the trunnion, to find the trunnion's axis. Whatever your NET valve lift is, then HALF of that should be the amount which the trunnion sits below the ROLLER PIN'S AXIS. By measuring the amount which the valve tip extends above the retainer's top, then dividing the roller in half and adding it to this, you get the **STACK HEIGHT** (height of the roller pin axis above the retainer). As shown above, the **ROCKER HEIGHT** is the SUM of the stack height and whatever the dimension of the trunnion axis is above or below the retainer's top plane.

NOTE: This illustration makes a reference to the TIP HEIGHT, showing it (logically) above the valve spring retainer. But this term is more specifically used to define the tip length of the valve. (See: p.48)

On **STAND MOUNTED SYSTEMS** (a.k.a. "shaft systems"), the same reference between the SHAFT Centerline and the ROLLER TIP Centerline is relevant. HOWEVER: With differing Intake and Exhaust valve lifts tied together by a common shaft, individually moving the rocker height as shown above is not possible, but each still require adjusting for their individual necessary to avoid wasted motion with one rocker arm or the other. So what do you do?

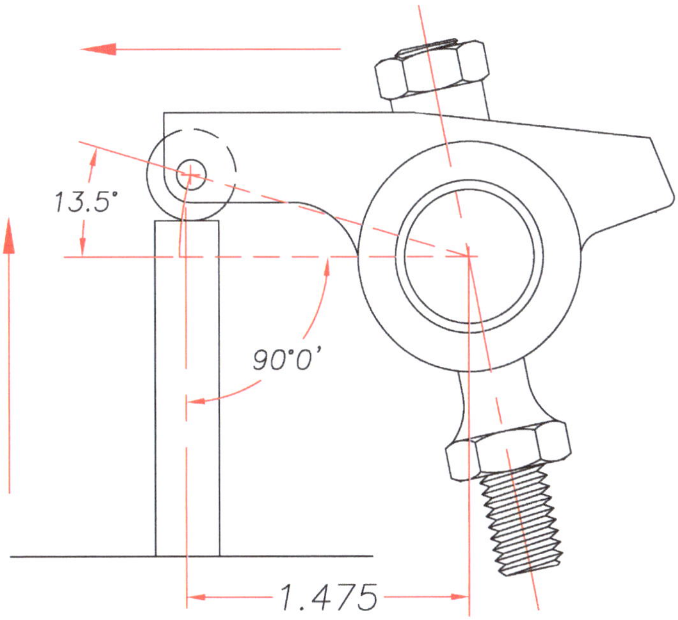

CONVENTIONAL
ROLLER ROCKER ARM
AT CLOSED VALVE POSITION

MOTION LINE

With the valve spring and all the measuring lines removed, we can see the **real line** that is important. The RED DASHED LINE running between the trunnion (or shaft) center and the roller pin axis is what we've always referred to as the "**motion line**." This line is what you are really adjusting the angle of when raising and lowering the rocker body in relation to the valve tip.

What is important to see here, is the 11-1/3° angle between the STUD and the VALVE. As the two white arrows depict, the more you increase the valve tip height (upward), the shorter the distance between the stud and valve becomes, so the more the rocker arm's roller will move across the top of the valve when sitting in the closed valve position as shown here. What is important is the *angle* of this *motion line* when the valve is closed. But we measure this by "height." If you follow the bad advice of those who tell you that correct geometry is placing the roller in the middle of the valve, then this angle changes based on varying valve tip and rocker design combinations that are impossible for any component manufacturer to predict or control, unless they make everything. The above 1.475" roller to trunnion length will yield .689" Valve Lift with a 27° rotation (13.5° Mid-Lift).

It makes no difference where the roller sits atop the valve*, as long as its wear pattern is on the valve. What matters is the "angle of motion" as the rocker moves this MOTION LINE (rocker loads) from closed valve, to mid-lift, to full-lift. As long as the angle when closed begins with half of the angle of motion (i.e., 13.5*° to MID-LIFT for 27° of full motion, as shown above), then these loads will balance in a straight down direction upon the valve, and NOT be over-arcing and shoving the valve against the guide, causing excessive friction, heat, wear *and horsepower loss.*

***OVER-ARCING** is the MOST COMMON symptom of rocker arm geometry found on most engines. There are many reasons for this, but simply, it's because few people understand how much and why the rocker needs to be RAISED to adjust its pivot points. The junk plastic pushrod checkers that have been sold for years, do NOTHING to set geometry accurately. In some cases, only making it worse. At the very least, they give the engine builder (and his or her customer) a very false sense of security.*

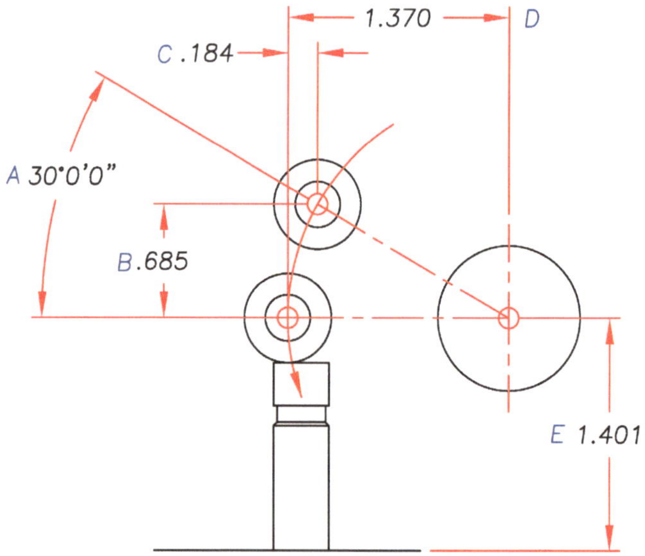

30° Radial Motion = .685" Valve Lift
ACCELERATES SLOW, THEN INCREASES
TOWARD FULL OPEN

OVER-ARCING

This Illustration shows OVER-ARCING, whereby the roller tip side of the rocker arm is moving outward and "OVER" its path as it pushes downward upon the valve. NOTE that "C" (ROLLER SWEEP), is **.184"**. The valve tip width is only .342" (11/32), so here more than 1/2 of the tip's width sees roll across action, which induces side loads on the valve tip, valve guide and valve stem. The roller does NOT "roll," much to the argument of many. It merely provides a "pivotal" axis for converting the linear motion into radial motion. In fact, anyone who's ever seen high speed photography of the roller working atop the valve, will confirm that is does more "skipping" than anything (kind of like an airplane tire first hitting the runway). Note the "Motion Lines" angle between the trunnion and roller axis.

Of the various points of information shown above, note "**A**" illustrates that 30° of rocker ROTATION is occurring to initiate "**B**", a valve lift of .685". (See: MID-LIFT ARC, p.22) This is wasted motion, compared to the valve's INSTALLED Geometry being set for MID-LIFT, which yields .**709**" VALVE LIFT for the SAME 30° of rocker rotation!

As with the UNDER-ARCING illustration, shown as the exact opposite perspective, both extremes illustrate how the "symptom" of a NET valve lift can be the same while opposite circumstances prevail. The main thing to understand from this aside from the obvious increase in roller SWEEP, is that the **valve lift speed is inverted** between *over-arcing* and *under-arcing*. The above has the valve starting off the seat slower, then increasing as the roller begins rolls across the valve to open it and begins to reach a 90 degree relationship, where peak radial velocity occurs. The speed of the valve's opening increases proportionately until reaching this 90 degree point. The crankshaft has had to turn many additional crank degrees to open the valve any specific amount compared to the direct pushing down that MID-LIFT geometry would have provided. The effect, is that the rocker arm has re-tuned the camshaft, diluting it.

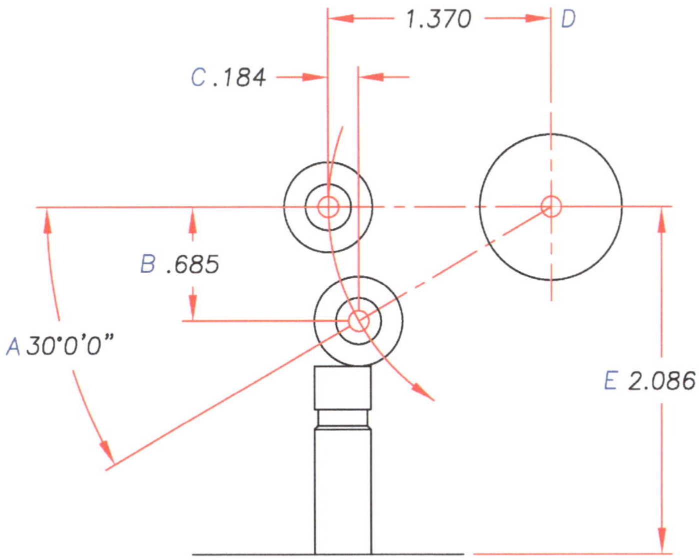

30° Radial Motion = .685" Valve Lift
ACCELERATES FAST, THEN DECREASES
TOWARD FULL OPEN

UNDER-ARCING

The above Illustration shows **UNDER-ARCING**, whereby the roller tip side of the rocker arm is moving inward and "UNDER" its path as it pushes downward upon the valve. Note the lower "Motion Line" angle of load at FULL LIFT (30° down), between the trunnion and roller axis, and consider the effect to the valve guide. As with the **OVER-ARCING** illustration, which is purposely shown as the exact opposite perspective, both extremes illustrate how the "symptom" of a NET valve lift can be the same, while opposite circumstances prevail. The main thing to understand from this, aside from the obvious increase in roller SWEEP over the **MID-LIFT ARC** example, is that the valve lift speed is inverted between over-arcing and under-arcing.

In this example, the valve is starting off the seat fast, then slowing as the roller begins to leave the linear path pushing down upon the valve. When it begins to follow under its axis, as shown above, the speed of the valve's opening slows proportionately. Again, just as with the previous example, this is measured at the crankshaft as additional degrees of wasted rotation, compared to **MID-LIFT** geometry.

Remember, these illustrations are shown only as a "snap shot" of full valve lift, but the effect on additional crank rotation is throughout the lift cycle, requiring the crank to turn further for each increment of valve lift. But it is more detrimental than that, because it is wasting all THREE Cam Dynamics: LIFT, DURATION and VELOCITY from what the cam is really designed to implement. It's also worth noting, that like "ratio" (also measured only as a snap shot taken at full lift), no one ever checks these effects in between closed valve and full open valve lift; unless you've got a Cam Doctor or go through the extra time to graph this acceleration on paper in fine valve lift and crank degree increments. So you never appreciate these losses.

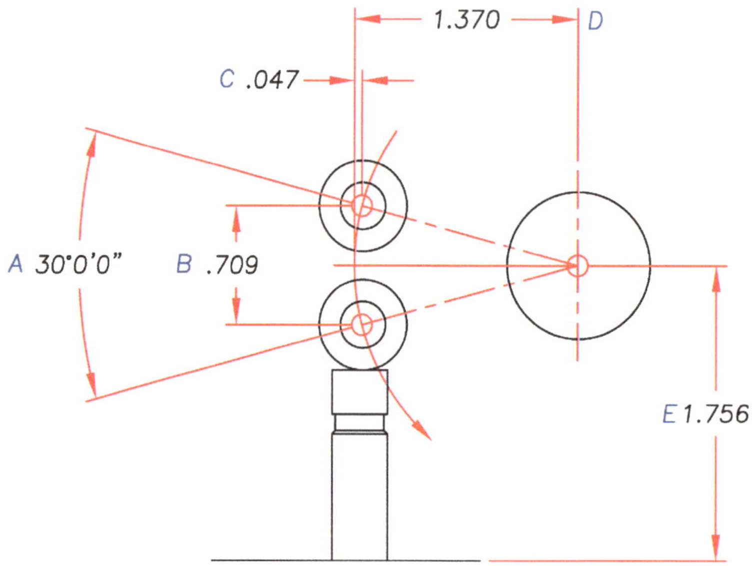

MID-LIFT ARCING

When the valve's INSTALLED Geometry has been set correctly, the maximum amount of *balanced* angular motion is transferred to the valve. The little TRIANGLE created by the two "Motion Lines" (p.172) represents the "balanced" load from zero lift to full lift upon the valve guide. In this case, it is **.709"** for 30° rocker rotation. The **OVER-ARCING** and **UNDER-ARCING** examples in the preceding pages depicted the NET Valve Lift was significantly less at only .685". This loss was a "symptom." The rocker's rotation was still 30°, but the net down motion effected to the valve's linear path was reduced by **.024" Net Lift**. This is because the roller was exceeding the minimum linear path and following itself either over or under the axis of rotation. Historically, rocker companies ignored this and chose to inject additional correction to the adjusting screw for making up what they perceived was lost ratio. Bad news.

If you compare this .709" over .685" as a percentage, you will see the "efficiency" has increased 3.4%. If the DURATION of cam event also increased 3.4%, a 270° cam would now yield 279°. *Is that increase worth using MID-LIFT geometry?* Another way to look at this, is NOT running MID-LIFT geometry means that you've LOST 9° from your cam. Sadly, this happens in top shops every day.

NOTE "**C**", the ROLLER SWEEP in our MID-LIFT example above. It is **.047"**. As you've seen on the preceding examples, they were **.184"** across the valve. This too is a symptom. Granted, it is a "predictable" symptom for those using a computer or draft table accurately. But it is NOT the best way to set ROCKER Geometry, as one tool being sold does. I will admit that it is better than plastic pushrod checkers, but if falls way short of the accuracy needed to maximize cam efficiency at the valve. The rocker height Reference (**E**) is shown to be in the middle of what was needed, from the OVER-ARCING and UNDER-ARCING illustrations, while the Rocker Length of 1.370" remains the same. What is NOT shown on any of these, though, is the effect of raising and lowering the rocker arm, or extending your valve lengths in relation to the stand or stud mounting.

~jM

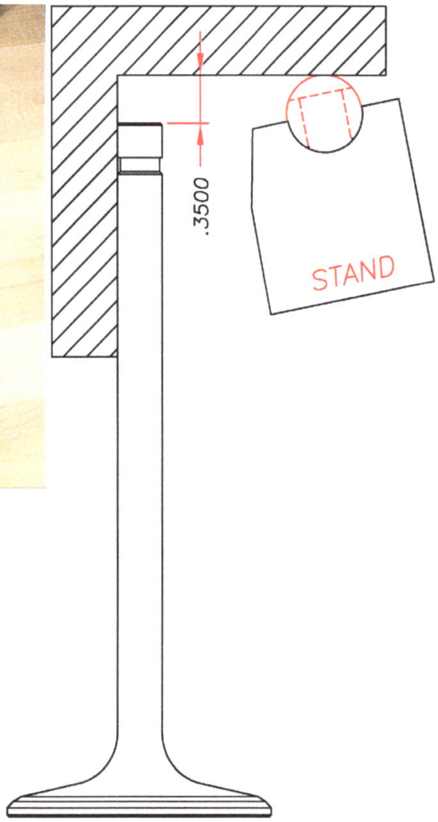

INSTALLED GEOMETRY™

STAND MOUNT ROCKER SYSTEMS

Precise STAND HEIGHT SETTING for ANY Shaft Rocker System of ANY Manufacturer

STAND MOUNT ROCKER SYSTEMS
INTRODUCTION

STAND MOUNT ROCKER SYSTEMS (a.k.a. "shaft rockers") are considered the top choice in valve-trains where roller cams and high RPM engines require high valve spring pressures. Many companies have come and gone, and a few of the more well known have continued to be with us for quite some time. They couldn't really survive, let alone prosper if they didn't make a reasonably reliable product to endure their extreme use. Of the top name companies, each have their own story of how their rocker arm designs evolved. Going back to examine their early offerings explains many things, and even with the biggest name brands, you'll see considerable back and forth experimenting with size, shapes, materials and dimensions, including the very geometry principles they were designed by, or – as in most cases – completely ignored.

Chrysler Racing Engineers didn't <u>mandate</u> their 1999 teams use MID-LIFT geometry for nothing; nor did General Motors publish MID-LIFT instructions as their high performance recommendations in 1992, if there wasn't something well founded in the MID-LIFT formula. In spite of this, I've had to spend the better part of 30 years explaining as much about the reasons why various cam and rocker companies still promote wrong geometry or no geometry, as I have explaining what MID-LIFT geometry is. To fix problems with your existing system you can refer to the chapter on "DESGIN Geometry." This addresses checking and changing the adjusting screw position, which as you will learn, is a critical component of geometry. Don't expect the manufacturer to tell you.

Good pushrods aren't cheap. Good cams aren't cheap. Yet the rhetoric of how to set up rocker arms on an engine has consequences to how that cam will work with your heads, and what length pushrods you'll need. If you have the wrong information, both of these items are likely to be the wrong parts for you to get the most out of your heads and avoid excessive wear and tear.

The following is simple. There are numerous easy ways to measure what is needed for setting rocker geometry correctly. Some involve tools, some measure from MID-LIFT, some from CLOSED valve. This is just one of several ways to use simple tools that any good engine builder will possess to precisely set the rockers for the most efficient operation with the valve. This provides the least amount of wasted motion, the least amount of side loads, and the least amount of wear and harmonic acceleration. When you know how important this is, the biggest problem you're likely to have is wondering why some of these companies you've trusted never filled you in long ago. See: *The History Of MID-LIFT™*. The reason isn't as sinister as it sounds.

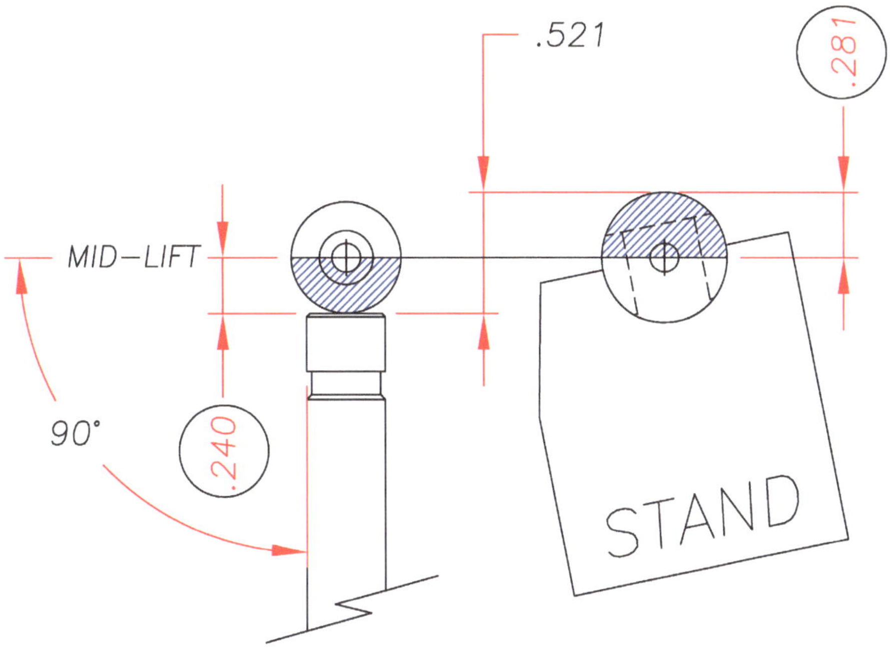

The above is what we are trying to set when the valve is at MID-LIFT. The rocker body is stripped away and only the relevant parts are shown. These are: the STAND; the SHAFT (end view), the VALVE TIP, and the rocker's ROLLER. This is shown as if the valve was open to half its net lift, but we don't need to move the valve to MID-LIFT to achieve this; we will set everything from a CLOSED valve (zero lift), and be much more precise in the process.

Please note the two half-round crosshatch patterns. These are the BOTTOM half of the ROLLER, and the TOP half of the ROCKER SHAFT. When the rocker's roller axis has reached a 90° angle with the valve centerline – *this* is their relationship to each other.

Gauges and tools that would compensate for diameters can be easily made, but for an engine builder setting the stand height for the exact MID-LIFT Installed Geometry, it is easiest to simply measure the external diameters with the valve tip from a closed valve position. To convert diameters to centerline locations we need to use the radii of these diameters shown by the cross hatches. We ADD them together, and this gives us a STANDARD formula that is unique for this rocker system, dictated solely by the diameters of these two parts. Roller diameters vary from one manufacturer to another; accordingly, so too will this standard. In this PRO-STAND™ example, the .281" shaft radius and .240" roller radius (encircled dimensions) total up to .521".

On your brand of system, you need to do the same thing. Measure your rocker's roller diameter; then measure the shaft (you'll likely have a 9/16" or 5/8"); then add HALF their values together.

Whatever these two diameters are for YOUR system, will be the FORMULA, so be sure to <u>write this down</u>. **THIS IS STEP #1.**

ADVANCED ROCKER ARM TECH

Jim Miller

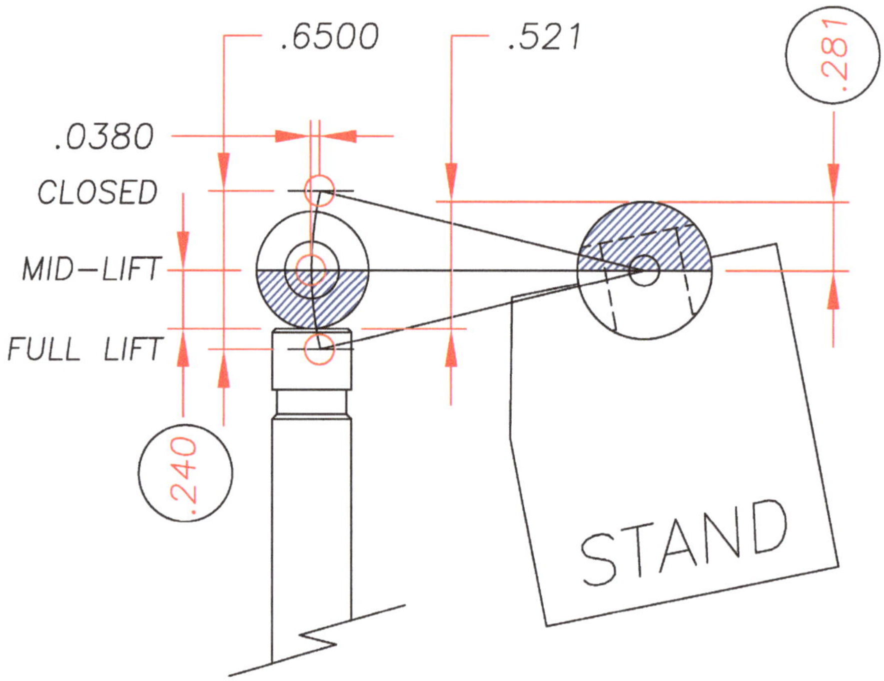

You **MUST KNOW YOUR NET VALVE LIFT**.

It is IMPOSSIBLE to set rocker geometry without knowing what the net valve lift is. Any tool, or technique that tells you what to set your stand or pushrod length to, without taking NET valve lift into consideration, is a complete waste of time that is misleading you. In the EXAMPLE here, I've arbitrarily chose .650" as our NET VALVE LIFT. Whatever your NET VL is, will be **STEP #2.** Write it down!

For illustration purposes, the previous view is shown above with the addition of the .650" datum circles to depict CLOSED valve, MID-LIFT and FULL LIFT positions. As before, this represents positions with valve at MID-LIFT. Note the in-and-out sweep of the datum circles (witness mark on valve tip) is only .038". Not shown here is a length from the shaft to the roller of 1.410", which is the Miller Pro-Stand™ small block Ford and Chevrolet system rocker length. A big block rocker's sweep would be even less.

FIVE STEPS:

1. **Add your rocker shaft and roller tip diameters, then divide in two for your FORMULA.**
2. **Measure TOP of SHAFT over VALVE TIP when Valve is CLOSED (See: p.23).**
3. **Subtract 1/2 of your NET Valve Lift from your particular FORMULA (i.e. .521").**
4. **Subtract the SUM of #3 from #2 (above).**
5. **Add 2% of your SUM for #4 to that sum to allow for INCLINED stand angle.***
   (* SB and BB Ford, Chevy and other engines with positive inclined Stand Angle of 10° or 11°.)

The FINAL FIGURE you have from Line 5, is how much the stand needs to be moved. If it is a positive number as our example shows, then you would machine the stand that amount. If it is a negative number, then you would SHIM the stand to what is needed. As you look through the following illustrations you will see everything amounts to only these FIVE steps in determining how much you need to change your stand height for proper geometry.

ADVANCED ROCKER ARM TECH

Jim Miller

The square resting on the top diameter of the shaft is shown referencing along valve stem, making BENCH measurements easy.

For checks on fully assembled heads, or on the engine, use the Valve Spring diameter. The valve spring RETAINER'S top surface can be used with a straight edge and creative use of feeler gauges.

The key is to be sure you are 90° to the valve stem when establishing the shaft height above the stem tip.

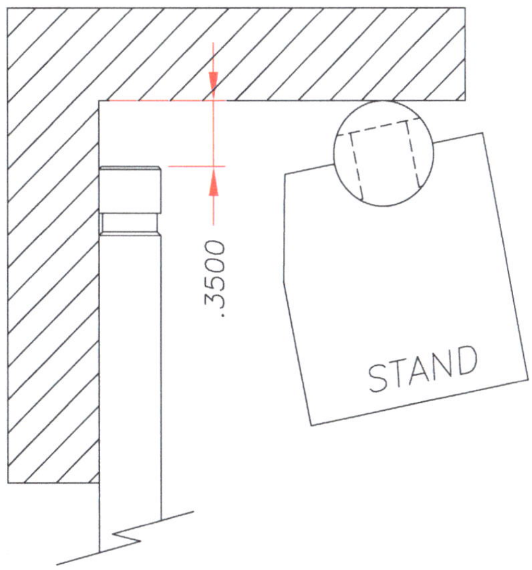

After establishing the FORMULA your rockers have, the second thing that must be done is to measure your VALVE TIP'S height with the height of your rocker system's shaft, when the valve is CLOSED. They KEY here is to do it at a 90° angle with the valve, as shown.

You do not need to have the heads on the engine, nor do you need to have the heads assembled. You merely have to have the valves installed and closed, and you need your rocker system's stands mounted to the head with a couple of bolts, just to secure it from moving; then lay a bare shaft in place without rockers.

A machinist SQUARE is then laid along the side of the valve and across the top of the rocker shaft. Depending on many things, you'll have some space similar to what is shown above. It could be .100" or it could be .500" or any figure in between. Don't worry about what it is for now, we just want to measure this and write it down on with the formula you recorded.

> **NOTE (1):** The MILLER PRO-STAND™ systems were always supplied with a surplus amount of material on the bottom to machine for final assembly, and thus avoid "shimming." With most other "shaft system" manufacturers who rely on "SHIMS," the above measurement with their stands may likely have given a much smaller value, like .100" or .150", and when you get to the final formulas for setting your stand height, you will likely need to SHIM "up" to get the correct geometry, unless you've been using too many shims already. Then you'll be removing them. This is a good thing.
>
> **NOTE (2):** *For engines with TWO rockers mounted on a COMMON shaft such as SB Chevy, if they have different valve lifts it is important that you use the valve with the greatest lift, as the stand will sit lower. Secondly, where valve lift is different between the exhaust and intake of a common shaft, the valve with the greater lift will ideally have its TIP taller by 50% of the Valve Lift DIFFERENCE. (i.e., Intake Valve .650" VL, and Exhaust Valve .600" VL = Intake Valve Tip will be .025" TALLER than the Exhaust.)*

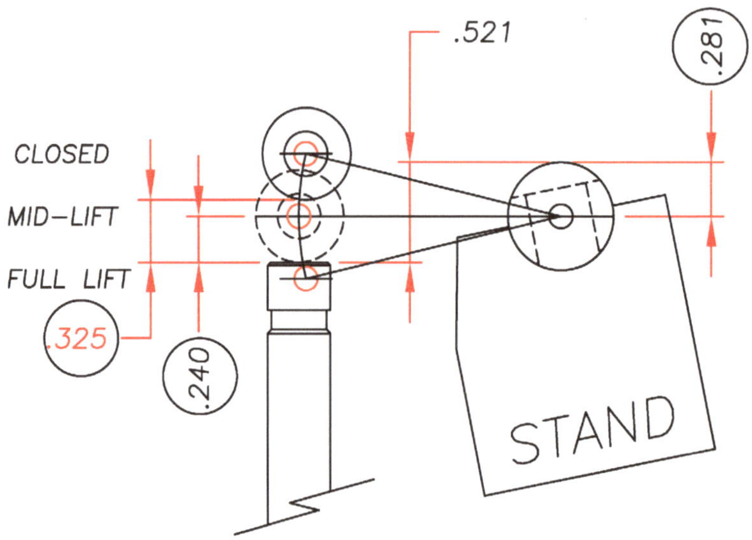

In this illustration the CLOSED valve roller (solid circle) is shown **.325"** above the MID-LIFT roller position (dashed lines).

What is key here is that we are measuring from the TIP of the valve to the BOTTOM of the roller. When the valve is CLOSED, that is where it NEEDS to be. Or more simply stated, subtracting that .325" from our FORMULA of .521", will give us the answer to how far BELOW the SHAFT the valve tip needs to be.

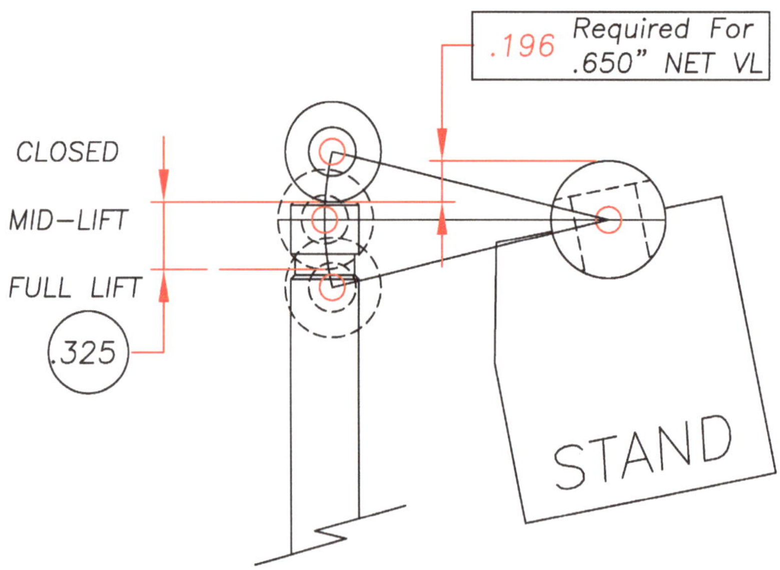

With the VALVE shown in the CLOSED position, we see **the .196" sum** derived from subtracting the .325" from the .521" formula.

To get the stand to have this final value, simply subtract the required .196" from the measured .350" shaft over valve height. That leaves **.154"** the stand will need to be **LOWERED** in this example.

All my stand mounted (shaft) systems were made with a surplus of material on the stand, which required the engine builder to fit and machine the stand as needed. Shims are a compromise; a cheap and convenient way out for both the manufacturer and the engine builder. But with [nearly] all shaft rocker manufacturers you will have to RAISE the stand with shims to get your final value.

~jM

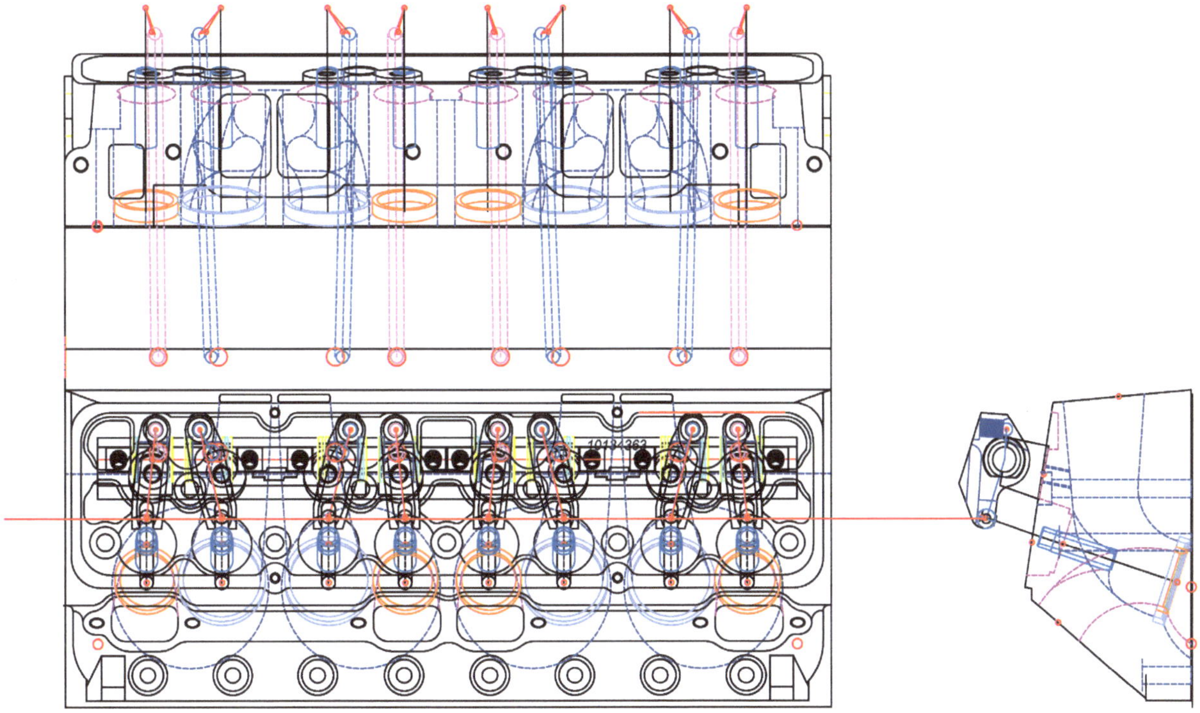

DESIGN GEOMETRY™

The Principles of Precision Geometry

MID-LIFT™

Rocker Design and Application

Note: This chapter may contain some repetition of points from other chapters. For continuity most of these have been edited wherever possible, but some have been left intact for emphasis where relevant.

DESIGN GEOMETRY™

INTRODUCTION

The following information pertains to the "**Design Geometry**" of a rocker arm for the Over Head Valve engines (non-over head cam). This is one half of the equation for an optimum valve-train on any such engine, but it is the most important half. The other half of this equation is called "**Installed Geometry**."

DESIGN Geometry is specific to the actual design angles of the pushrod with the valve on any given engine, and in most cases these are different from one brand and model of engine to another. This was the missing link long ago, which continues to be produced by top name cam and rocker companies to this day for a variety of reasons. Mainly the cost associated with changing an existing product line they've built their name around before MID-LIFT was developed and made public, and also to avoid tainting a name brand recognition that is built on the image of leading edge innovation and engineering. So, some companies continue to ignore, disregard and even refute MID-LIFT design technology is real, let alone better. However, keep in mind that GM's Race Engineers officially recommended it as far back as 1992, and Chrysler's Racing Division mandated it for the debut of all their NASCAR teams in 1999.

INSTALLED Geometry is what the engine builder "installing" the rocker system upon the engine determines, by fixing the height of the rocker's pivot points (trunnion or shaft) in relation to the tip of the valve. With stud mount engines this is done by changing the pushrod length, where the tail of the rocker rests. Because the pushrod is farther from the tip of the valve than the trunnion – which is the axis to be raised – the pushrod will be extended about 1.3 times more than this trunnion axis, depending on rocker ratio. The engine builder will set the trunnion axis exactly half the net valve lift BELOW the roller tip's axis at closed valve. This chapter features "Stud Mount Installed Geometry" and "Stand Mount Installed Geometry," both illustrate easy and precise ways to set this, and can also be found at the MID-LIFT.com web site.

WRAP-AROUND EFFECT

I am a fanatic about precision INSTALLED and DESIGN geometry. I am critical of the adjusting screw pushrod tip axis being 90° with the pushrod. However, unlike the valve side, there is a wrap-around effect that is tolerated by minor errors here. Because a pushrod is free to travel in and out with the rocker, the linear cam information actually wraps around the rocker's axis, so if the cup axis is off by 2° or 3°, the ratio and net lift won't really change. But HARMONICS will.

The information in this chapter covers specific design parameters so you can determine what you need in any given system, and how to modify or change what you're using to get the most from it. Correct geometry is NOT making *sure the wear pattern is in the middle of the valve at mid-lift.* This has *nothing* to do with setting correct rocker geometry. In fact, you can make your geometry worse with such techniques when you don't know what is really important.

This is NOT rocket science. This is A-B-C High School Trigonometry. It's simple, once you know what the principles are – and you know why *NOT having it* is so bad. Most of the energy is spent un-learning the wrong way, still taught by big, famous companies who missed it decades ago. If you care about a 270° cam being 270° at the valve, and not 258°, then you need to do it right. You need to split the rocker arcs evenly on BOTH sides of the rocker arm, and do it precisely at MID-LIFT. Here's how, and why.

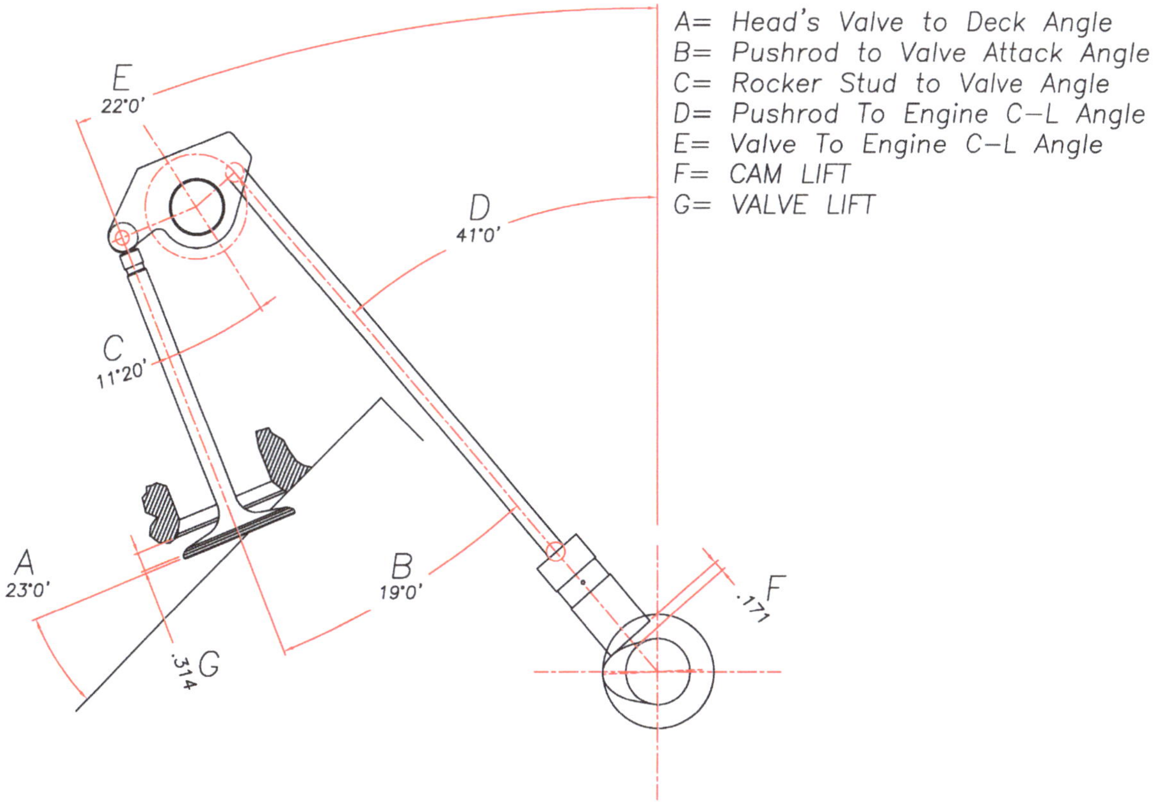

DESIGN GEOMETRY 101

STUD MOUNTED ROCKER ARMS

As noted in the opening chapters, the rocker arm's design geometry refers to the angle between the pushrod and valve centerlines. The **RED** circle around the rocker arm's axis represents the pushrod radial. This is the most critical side of the rocker arm in tracing the cam accurately. Here, it is shown with the pushrod's ball tip axis at a 90º relationship to the trunnion simultaneously when the valve centerline is 90º with the roller tip to trunnion axis. The inclined angle of the rocker arm's mounting stud is the SB Chevy's 11º20'. For every .100" the valve length is extended the rocker is moved across the valve tip by .020", which is why the roller's wear pattern atop the valve is not relative to correct rocker arm geometry.

Because roller tip rocker arms have historically been designed from measurements taken in the "closed valve" position, instead of the MID-LIFT position as shown above, there is a wide range of measurements between rocker brands and one engine to another. Accordingly, rocker ratios are not accurate, because the valve lift was not taken into account by designing in the closed position.

If MID-LIFT is used from the beginning everything balances. Both sides of the rocker arm are perpendicular so the very least amount of in-and-out motion occurs on BOTH sides. In spite of the above, very well known cam and rocker arm manufacturers actually promote an over-arcing geometry that imply it is better to reach a 90º relationship with the valve when it's closer to full lift (2/3 or so) because has less loading on the valve-train. That's an erroneous claim; it's malicious and negligent. The excessive back and forth motion of the pushrod and roller tip creates additional: flex; friction; side loads; and induces unnecessary harmonics that are deadly to all, especially the valve springs, forcing engine builders to run higher than needed spring pressure which compounds the problem.

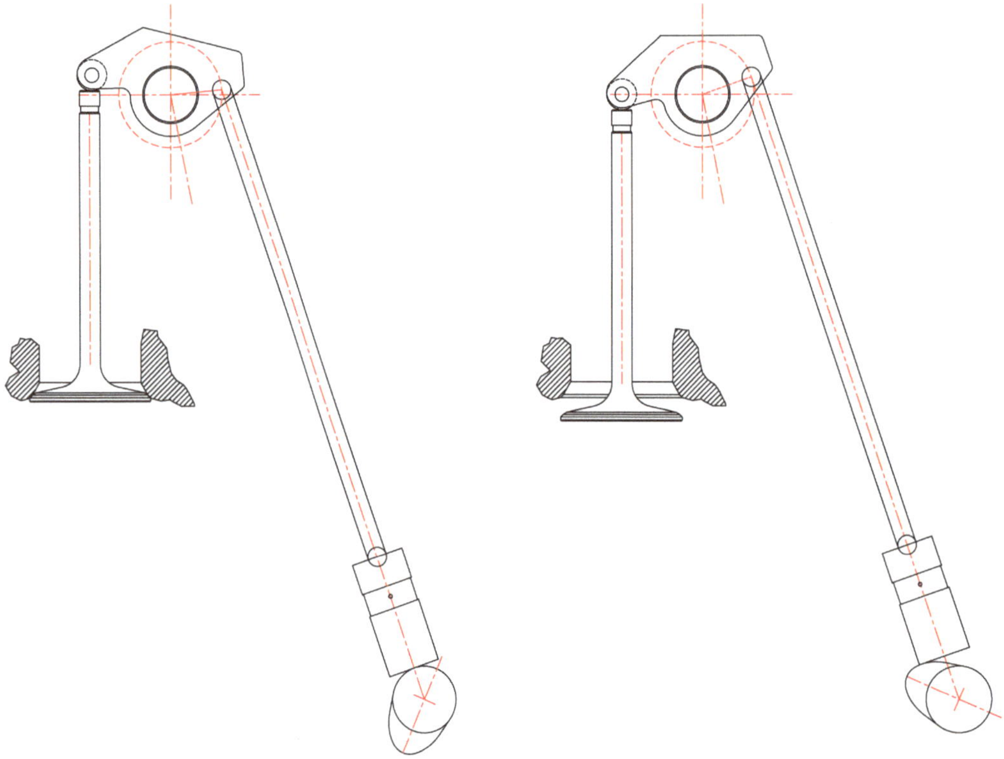

DESIGN GEOMETRY 101

The two views above show a PRO-STUD™ rocker arm silhouette in TWO stages: CLOSED VALVE and MID-LIFT. The **RED** dashed circle represents the "radial" element of the pushrod tip that converts "linear" information. This point is repeated throughout these chapters, but it can't be overstated.

How the industry has looked at rocker arm design with the engine by speaking only in linear terms has limited understanding of what's really happening; a problem that existed for too long. The camshaft is going to lift the tappet (cam follower) a specific precise value. If the center-to-center distance between the pushrod cup (or adjusting screw) to the rocker's axis is known then this cam lift is translated into a specific, predictable ANGLE of ROTATION. It's simple high school trigonometry, not rocket science. It gets complicated when this arc is randomly moved around; meaning: when it is *not* divided equally. A 90° relationship must be established at precisely MID-LIFT. You'll find this rule applied in all forms of mechanical engineering; *except the aftermarket roller rocker arm industry.*

The only way to do that is to AIM for the MID-LIFT position, from the CLOSED VALVE position. If this is done, then the SECOND HALF of cam motion beyond the MID-LIFT point will be duplicated in how the pushrod follows the arc around. The in and out motion that occurs, are SYMPTOMS. They are NOT something you try to design for and by. Making tools that set rocker geometry based on this, is clumsy and not accurate. It may be better than nothing, and it may be better than plastic pushrod checkers (which are a joke), but it is a very "loose" way to find MID-LIFT position. It is very easy to have the trunnion/shaft pivot height be off by .010" or .020" by only measuring the extended reach of the roller atop the valve. Measuring the in and out (horizontal) dimension when you're seeking the UP and DOWN (vertical) dimension is clumsy, inaccurate and inefficient.

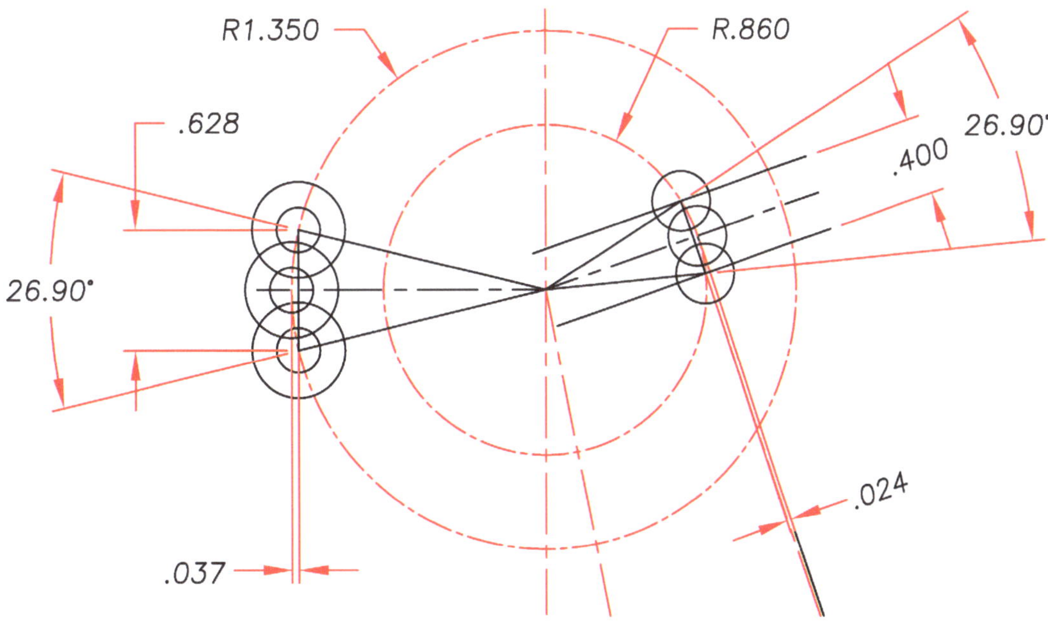

DESIGN GEOMETRY 102

The ABOVE is the essence of what matters with a rocker arm. With all rocker body features stripped away only the axis points of motion are represented. The right side depicts the *pushrod radial* of .860" at the SB Chevy's 19° pushrod to valve geometry. The left side depicts the *roller tip radial* of 1.350", and both are shown at perfect MID-LIFT *design* and *installed* geometry.

The 26.90° angle of rotation is dictated by the pushrod radial and cam lift, *but only* at Mid-Lift, where the above example of .400" cam lift generates a pushrod arc of only .024". Changing this balanced relationship of dividing the arc equally will increase the in-and-out motion will increase the harmonics of the pushrod, change the rocker's angle of rotation, reduce the duration of the cam, and *shift* the cam velocity and duration points with the crankshaft (area-under-the curve) from one spectrum of valve lift to the other, depending on which side of Mid-Lift the over-arc error was shifted to.

That same 26°90' angle of rotation occurs on the *valve tip* side to generate .628" valve lift, which produces a minimal .037" roller sweep across the valve, *but only* if it too is set to Mid-Lift geometry. Most important to note is not only this minimal sweep, but the *nature* of the sweep. The roller (as with the pushrod) each arc outward to their maximum point at *Mid-Lift*, then arc back to the same inward position as they began. On the valve side where the greatest linear motion occurs against spring loads working against the valve stem/guide, changing the valve tip geometry to anything less than Mid-Lift will increase the arc (roller motion) *across the valve*; increase the side loads imposed against the valve stem/guide; increase the generated heat, harmonics, and drag coefficients.

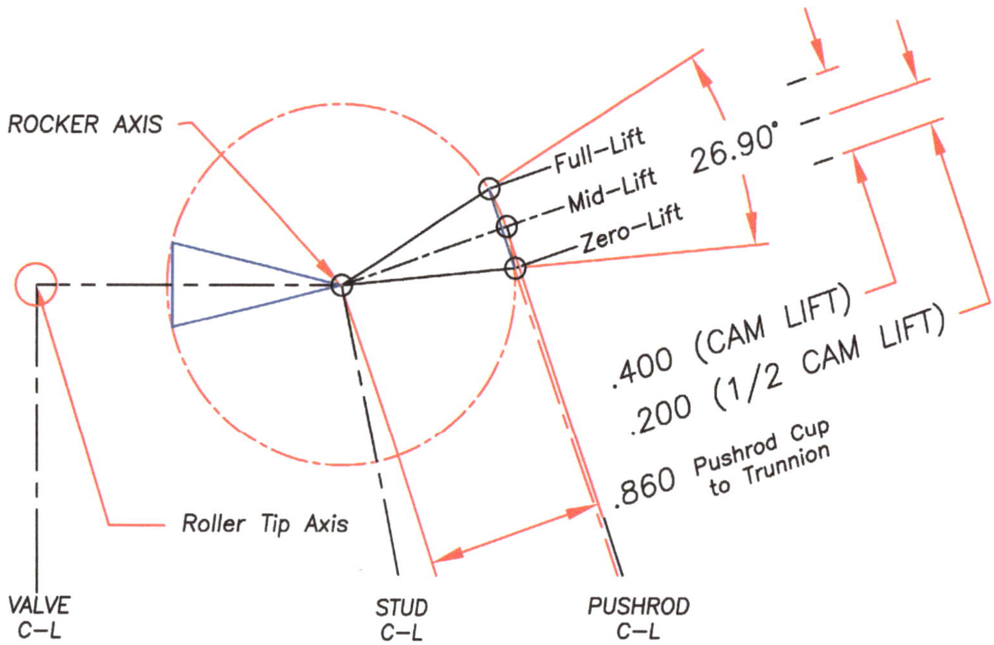

DESIGN GEOMETRY 103

The three small **BLACK** datum points (circles) depict the PUSH-ROD tip axis (*not tip end*).

The above represents the PUSHROD side of geometry. ONLY to the right half of our large red circle is relevant. This is the gospel that dictates everything. Here all three critical stages are shown, from the bottom up, with **BLACK** datum circles illustrating the **CLOSED** and **OPEN** points. The pushrod cup to trunnion centerline of the .860" radial creates an angular value from .400" cam lift of 26.90º.

It might be helpful to think of the RADIAL conversion taking place in the rocker arm as two forces: **DRIVING** and **DRIVEN**. The PUSHROD is DRIVING the rocker motion (as dictated by the CAM); and the VALVE is DRIVEN by whatever motion has already been determined by the rocker's DEGREES of ROTATION the cam imposed UPON the ROCKER. Why is this perspective important?

The **INSTALLED Geometry™** chapter (p.16) illustrated how OVER-ARCING and UNDER-ARCING are affected by 30° of rocker rotation, and how net valve lift was reduced from .709" (at MID-LIFT) to only .685". That **.024"** loss of lift was from the same 30° of rocker rotation! This was the DRIVEN SIDE.

NOTE: Digital Protractors are a good investment for checking your engine's pushrod to valve attack angles. For example, the standard 23º Small Block Chevy has a 19.0º Positive Pushrod to Valve angle. The 18º variant has a 14.0º pushrod to valve angle. A Big Block has an 18.0º Intake and a 20.5º Exhaust pushrod to valve angle (design geometry). This is what the three tangent points on the rocker need to be set to. For the better or worse; depending on the rocker arm manufacturer you're using, **rotating your adjusting screw up and down changes this significantly for just *one turn*.**

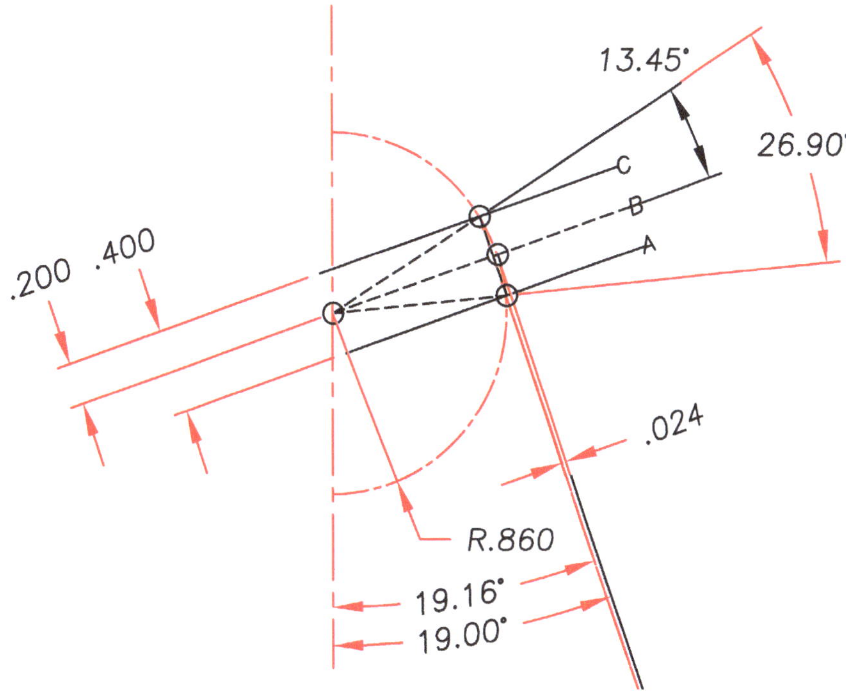

DESIGN GEOMETRY 104

The above is what I consider ideal. Unlike #103 of the preceding page, here we have only the tail end of the rocker datum points and their attached pushrod sweep of only 0.16° as they move from "A" to "B" then returning back to "C" at full lift, creating our 26.90° arc on a radial of .860" (pushrod cup distance).

Ideally we want is to FOLLOW the cam lobe as closely as possible through the TAPPET and the PUSHROD. The ONLY WAY to keep this radial motion to a minimum is to DIVIDE the arc into two equal halves, and to keep these other components as in-line with each other as possible. Setting the axis of motion to a 90° relationship exactly HALF-WAY through the stroke is the only way for the rocker arm to guarantee the very least amount of in-and-out motion and the very least amount of TRACKING deviation through the rocker arm. This is also "ratio deviation," and as stated previously, it includes all three CAM values: velocity, lift and duration; while the RATIO stays more constant throughout the entire lift curve. Since no one ever measures ratio outside of full lift, they never know what's being lost.

On the rocker arm's **DRIVING** side of LINEAR motion it is just the opposite. You can impose **MORE ANGULAR VELOCITY** into the rocker arm's radial arc by shifting the linear stroke to stop or start at the 90° point, or some value in between. In other words, be closer to 90° in CLOSED valve, or OPEN valve, instead of MID-LIFT. This will shift that tracking velocity of the cam from being at its highest wherever the 90° occurs. Moving the 90° point inverts the velocity of the cam, starting slow and accelerating, or accelerating and slowing, depending on whether the pushrod axis began at 90° or ended there. The following two pages show each of these extremes.

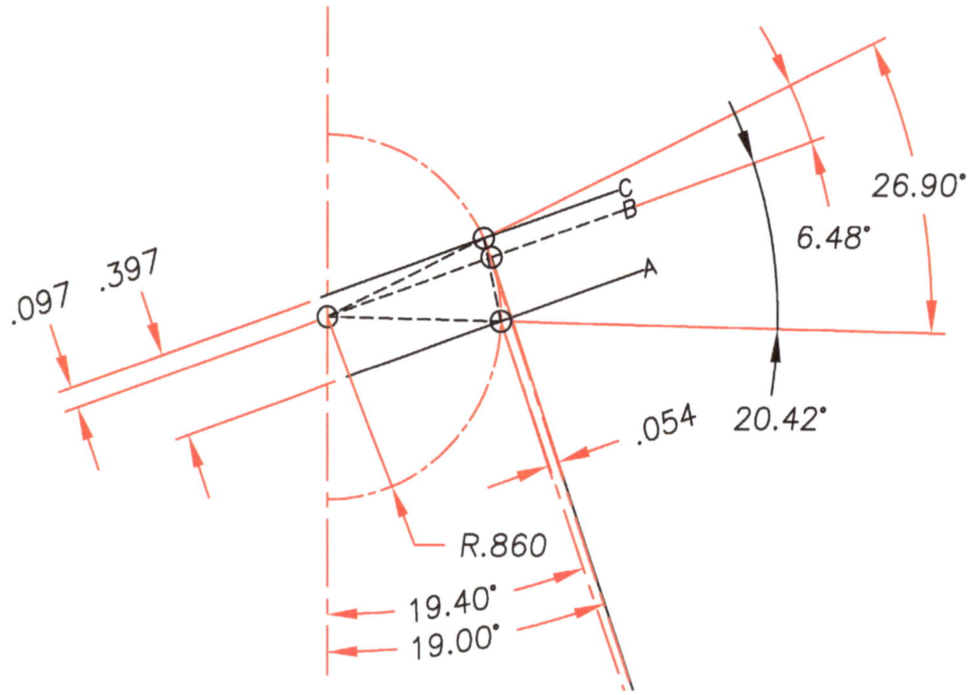

DESIGN GEOMETRY 105

This over-arcing pushrod is the geometrical effect of running a rocker arm with the pushrod cup .100" too LOW. Mid-Lift is delayed until .300" CAM LIFT (dashed line "**B**"), rotating the rocker arm 20.42º to reach this point. The peak cam lift of .397" is represented by solid line "**C**", only .097" beyond Mid-Lift, instead of .200" as it should be. The actual cam lift is still our .400", but .003" went *south*, deviating with the extra pushrod motion this over-arcing geometry created, and represented by the 19.40°; which is 0.24° more than the 0.16° MID-LIFT setting illustrated in **Design Geometry 104**, of the previous page.

However, all is not lost. Even though .003" is lost measuring perpendicular to the cam, another phenomena is going on which is unique ONLY to the pushrod side of the rocker. The *angle of rotation* remains the same. Because the pushrod tip motion is "fixed" to the cup, it follows to *wraps* around with the pushrod radial. So what? So this means the rocker arm still moves the same 26.90° as MID-LIFT. It is only when a more acute pushrod geometry occurs is this compromised. Also, notice the pushrod's in-and-out motion has gone from .024" to .054", *increasing by **125%***! Remember *harmonics?*

This shift in where MID-LIFT occurs has slowed "off the seat" velocity and accelerated the valve as it reaches peak lift. This is in addition to increasing harmonics with extra pushrod motion, to essentially change the cam's entire velocity and duration profile. Cam tests are meaningless in this state.

I noted that the angle of rotation remained the same *only* on this side of the rocker. If you go back to **INSTALLED GEOMETRY** (pp.20-22), you'll see what happens when the same angle or rotation is less than divided equally by MID-LIFT geometry. *Valve lift* is lost. *Velocity* is lost. *Duration* is lost. That's all.

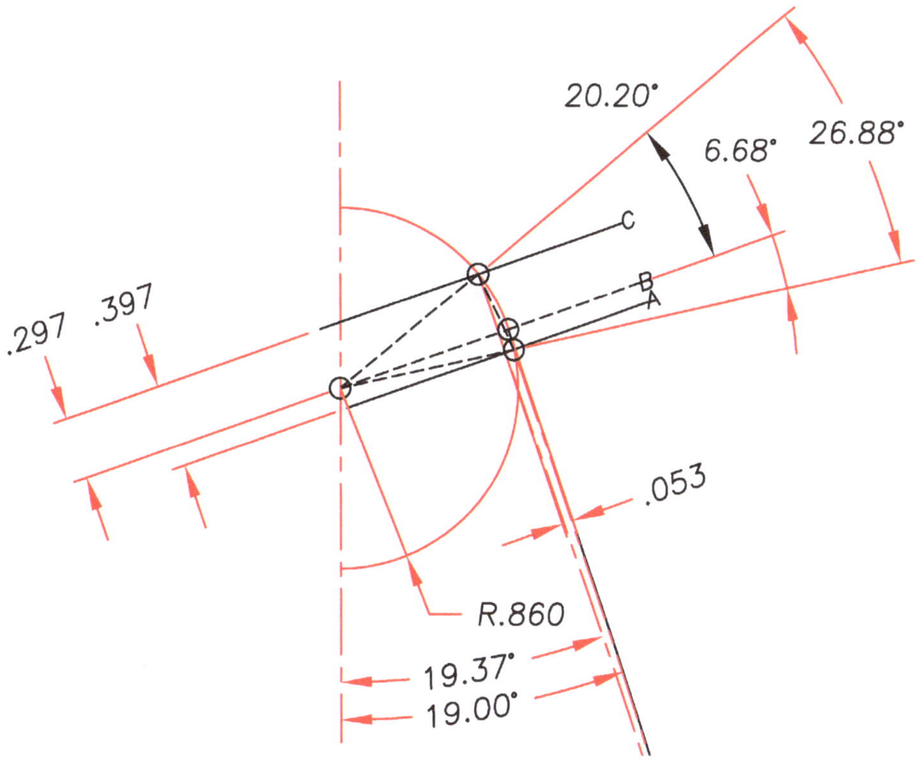

DESIGN GEOMETRY 106

Here the opposite extreme exists from the previous over-arcing scenario in **Design Geometry 105**, where the same .100" offset is flipped so MID-LIFT ("**B**") is reached *early*, as shown only .100" above the closed valve starting point ("**A**").

Because the deviation at full lift ("**C**") trails to the inside instead of the outside of the rocker's axis, we see a slight difference in the pushrod side motion and angle deviation. Instead of the .054" in the previous example, here it is .053"; and the pushrod angle deviation of 0.40° is reduced to 0.37°.

These deviations are not important, except as indicators of other symptoms going on. What this is indicative of is the effect each scenario have on the *CAM TIMING* translation through the rocker to the valve. The velocity shift from these two examples is extreme, and continues on to the valve.

Everything noted in the previous example is now reversed, where the rocker will now accelerate quickly in the first quarter of cam lift and then slow down as it reaches full lift. The downside to this is that when the rocker closes it will be reversing this process, to accelerate the valve closing as it reaches the valve seat, which is exactly what you do NOT what you want to do to the valve.

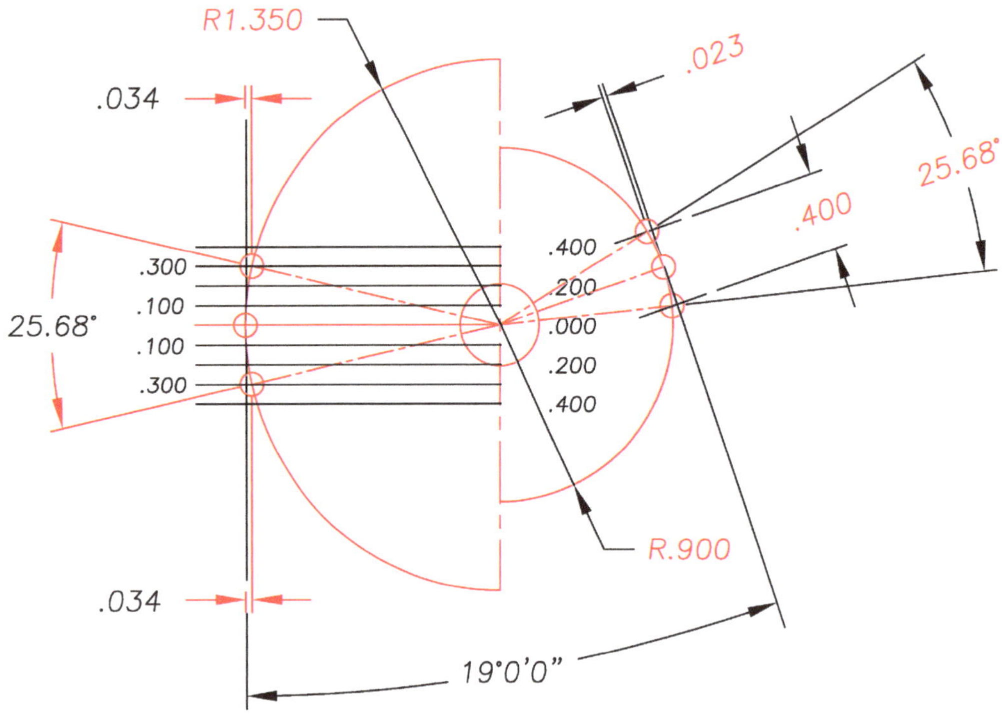

DESIGN GEOMETRY 107

Setting the *INSTALLED geometry* on the VALVE side of the rocker is the most accurate way to set the PUSHORD geometry on the right side, providing the rocker's **Design Geometry** matches the engine's pushrod attack angle to the valve, which in our illustration above is depicted at the bottom as 19°. This accuracy is **critical**. Keep in mind that only the distance between CLOSED valve and MID-LIFT matters. (Forget about the full CAM lift or VALVE lift.)

If the camshaft has .400" lobe lift (as above), we only seek to intercept the .200" cam lift point to make the pushrod axis 90º with the rocker. Whatever amount we "miss" on the first half of opening is thrown onto the back half of the cam lift. In the above, if the .200" is missed by .050" and our 90° relationship is reached at .150" (a too long pushrod), then there will be .250" of motion *beyond Mid-Lift* on the second half of cam lift. This would accelerate the rocker velocity quicker off the valve seat, but *slow it down* on the second half of the lifting cycle. On closing this would REVERSE to speed up the valve's closing.

This extra .050" can affect your valve timing with the crank by several degrees. If the opposite syndrome occurs by the 90º being established when the cam lift is at .250" (too short a pushrod), then this will make the valve react slower in the first half of the lift cycle, and speed up on the second half. And once again, when the valve closes it REVERSES this pattern. Both of these symptoms are effects that change the area under the curve (the active valve lift curve determined by how quickly it opens).

DESIGN GEOMETRY 201

STAND MOUNTED ROCKER ARMS

Although I have refrained from explaining this in great detail over the years, the rocker arm's tail (adjusting screw or pushrod cup) is the essence of Design Geometry because the pushrod side is the beginning of everything. Stud mounted rocker arms with their fixed pushrod cup design have their design geometry set by the rocker manufacturer, and all you can do is pivot the trunnion as needed to get your "Installed Geometry" where you want for the valve side by setting the length of your pushrod.

Stand mounted rocker arms (a.k.a. "shaft systems") are more unique and quite tricky because of their adjusting screw to set lash. You'll see in the following illustrations that their efficiency is totally controlled by the engine builder; but ironically, the rocker manufacturer (whomever that may be) never, ever gives any instructions about using the adjusting screw for more than just setting lash. I began the process of using the adjusting screw as a guide for its position in 1980, then making the adjusting nut a specific height to match the top so if it was out too far it was obvious. Yes, I'm the guy that began using the term "one and a half to two threads out." Years later the term was copied by a couple of notables, but even this was only explained as what's needed to avoid weakening the adjuster, and not for setting geometry.

Fact of the matter is, the adjusting screw is quite critical in its effect with the rocker arm's performance, and your cam's getting the information through to the valve. By now, you've already seen how specific the little red circles are placed to be "divided equally" in the foregoing pages, where the rocker arms were stripped away and only the tangents and motion line were displayed. That was to hopefully remove some of the visual clutter about rocker arms, which various shaped rocker bodies can impose on the mind trying to learn something new, and especially so simple.

The following pages will illustrate the rocker body on a variety of points, and show you the effect of variations between MID-LIFT geometry and the rhetoric of the most well known names in shaft mount rockers. One of the top names in shaft systems still insists their geometry is the best, which places the roller 90º when the valve is about 3/4 of the way through its lift. In most of these rocker designs the adjusting screw is placed at a disadvantage angle, because it was designed to line up with the pushrod when the valve is closed. Original thinking was this was best for setting lash. Later designs began to tilt some, but not enough, and usually for no specific reason. This has been a consequence of companies designing rocker arms by only measuring the head on a bench, and not the engine.

The final illustrations depict an example rocker with an adjusting screw that more closely matches the inherent philosophy of the top brand (well advertised) rocker arm that has adorned championship cars for many years. Race teams have won many races with rocker systems that are far from precise or efficient, and likely continue to do so. But the fact of the matter is, most of the championship users of this well-known system do NOT run it by the instructions the company promotes. The top teams modify the rocker to MID-LIFT geometry. They don't advertise this, as it was a well-guarded secret. But I'm illustrating it here, much as I did for the top wrenches of NASCAR 15 years ago; to show how important these little turns of the adjusting screw are – to get exactly what the cam can give, no more and no less.

The last thing you want to do is use rocker arms as a second "cam" by inducing acceleration tricks to the valve with over-arcing geometry; or placing the 90º tangent points anywhere other than MID-LIFT. Because when the rocker returns in the opposite direction, whatever you wanted to happen on the up side is now happening in reverse on the down side. Anything other than MID-LIFT is bad news. Period.

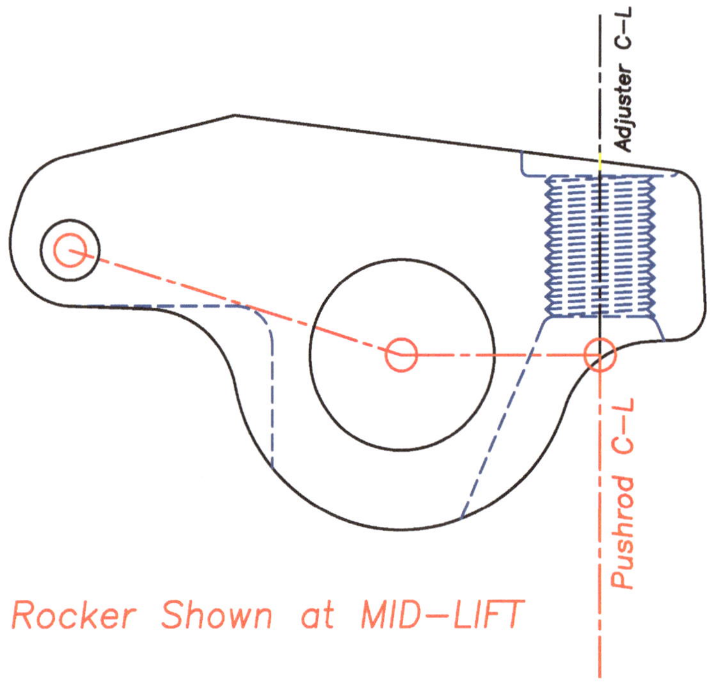

DESIGN GEOMETRY 201

Here's the standard type stand mount rocker silhouette, with adjuster threads, pushrod relief and valve spring relief shown in hidden lines (blue); absent the adjusting screw, but showing the adjusting crew Center-Line (C-L). Not shown in following illustrations is the PUSHROD C-L, and as noted in **RED** – this illustration is how everything would relate when the Rocker/Valve was at exact MID-LIFT.

As with prior drawings the red circles depict the exact axis intersections that matter. The lines running between them are the same motion lines shown on previous illustrations without the rocker body, while the vertical **BLACK** dashed line is the adjusting screw centerline.

As can be seen, our pushrod axis tangent would be movable up and down with whatever position the adjusting screw is at, determined by the pushrod length after the stand has been mounted.

> **NOTE:** As in the illustration above, ALL of the FOLLOWING illustrations assume the Pushrod would be VERTICAL as depicted in **RED** here, except as noted otherwise. It has been removed for simplifying our focus on the tangents. They also assume to be in MID-LIFT, meaning the rocker would rotate equally in both directions for closed and open valve.

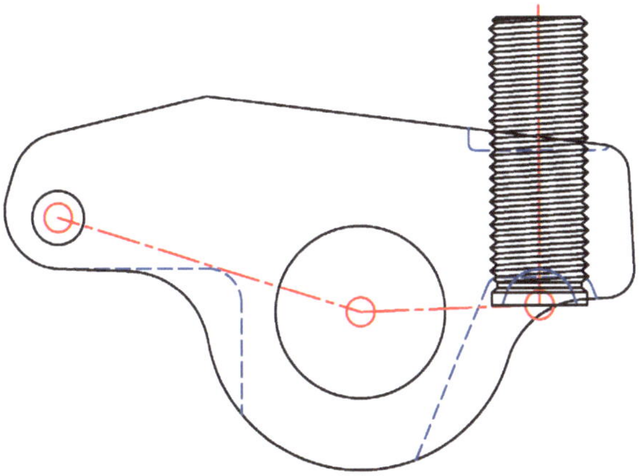

There are essentially two principle styles of adjusting screws.

The **CUP** style (left), is the most popular on most engines using shaft rockers, mainly because pushrod oiling is used.

Some cup depths vary, so you need to measure yours to determine if it is half the diameter of the pushrod ball (this example is). The pushrod's tangent AXIS is the BOTTOM SURFACE of the adjuster's cup. (The lock nut and washer were left out of view for clarity.) Good systems will have predictable length adjusters you can preset by the thread count to the top of nut. Only small changes are needed between 21°, 18° and 15° head designs.

The **BALL** adjuster (right) is the second principal means of setting lash on most stand mount rockers, and can be found more on older shaft rocker systems, although it is an option with most rocker arm manufacturers.

Typically the ball tips will be either 5/16" or 3/8" in diameter, so naturally the axis would be half this size. The ball tips were popular on several factory systems such as Ford FE (390-428); and Chryslers. The larger 3/8" ball is stronger and recommended if you have a choice.

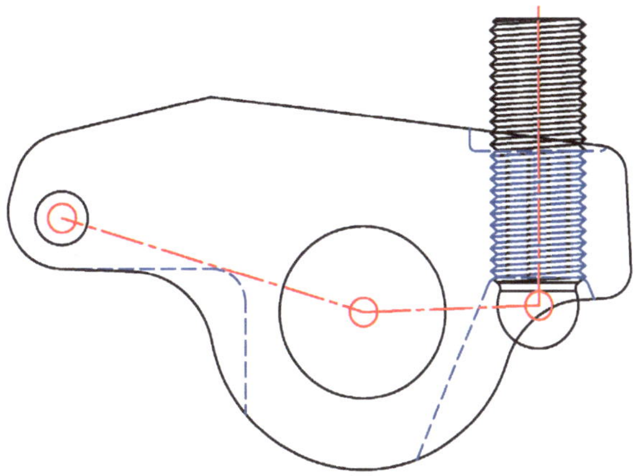

DESIGN GEOMETRY 201

Choosing between CUP and BALL adjusters may depend on the rocker body's design and if it may be weakened from increased machining for pushrod clearance. BALL adjusters of course require a CUP pushrod. The wall thickness of some CUP pushrod ends can require a larger pushrod relief in the rocker body over that of a CUP adjuster that uses a simpler ball-end pushrod. Proper MID-LIFT designs keep the adjuster in alignment with the pushrod's arc. But many "name brand" rocker arms have excessive over-arcing from an adjuster that is too high, and/or at an improper angle that fails to line up with the pushrod at MID-LIFT. The excessive side thrust is placed on the brim of the cup and is at its greatest angle when reaching maximum loads at full valve lift, so they fail. Spring loads are usually blamed, when in fact it is bad geometry, and the engine builder just didn't know what to look for.

Another factor, and usually the only one considered, is OIL delivery. Most BALL adjusters are SOLID and used only on rockers that have oiling other than pushrod; while the CUP is usually pushrod oiled.

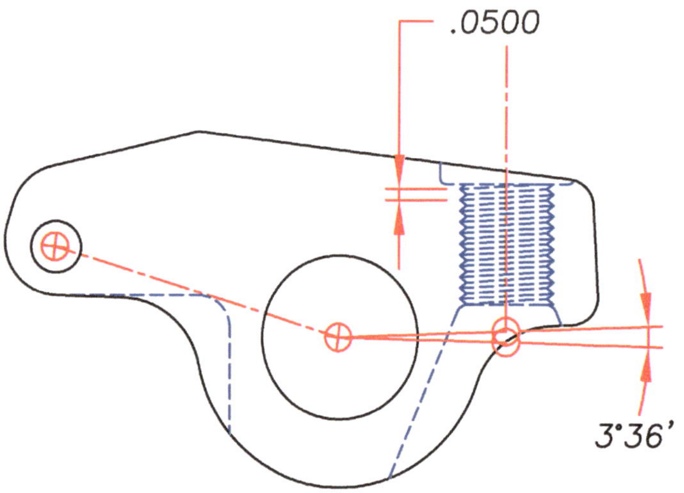

To the left, the 7/16-20 SAE threads have a pitch of .050" for each turn.

Some manufacturers use a 3/8"-24 SAE is .041" per turn. We'll round out to the 7/16", since it is stronger and the even value makes life simpler.

As can be seen by the second axis datum, one turn of the adjuster equals 3.5°. If you're trying to control rocker geometry to within 1.0°, then you cannot arbitrarily cut corners on custom pushrod lengths.

The highly popular Chevrolet SB head has a 23° valve to deck angle. The rocker to the right shows the amount of change required to extend the adjuster 5° for MID-LIFT geometry for the Chevrolet 18° SB.

Notice this 5° is only .069", which is barely 1/3 of a turn more than the above example. THIS is how precise you must be about the adjusting screw position. You lock it to what the design requires, then you use an adjustable pushrod to find the exact length needed to fill the space.

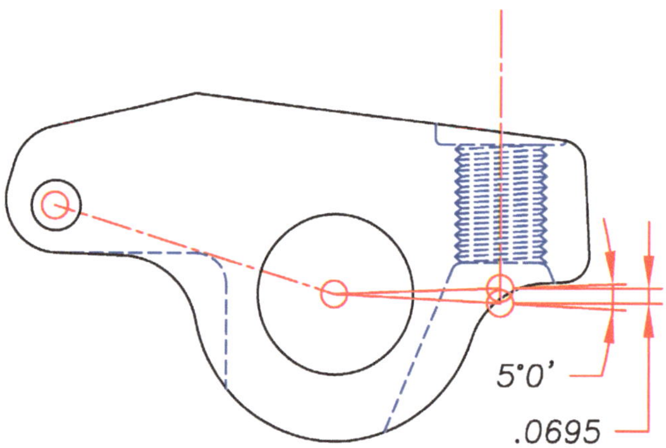

DESIGN GEOMETRY 202

The point to be made in the above examples is how sensitive the adjusting screw is to a level of precision that is necessary if you want to have optimum, maximum cam efficiency through the rocker arm to the valve. It all begins on this side of the rocker, and dividing the cam's lift precisely in half is critical if you want to get the ultimate area under the curve throughout the entire valve lift cycle. You can NOT be turning the adjusting screw one, two or three turns arbitrarily, even if your rocker manufacturer didn't design for such precision, and worse, failed to educate you to how important this is (as most – or all – fail to do).

To give some perspective as to how critical the adjusting screw position (or fixed cup in a stud rocker) really is, just for even numbers let's consider that your cam lift will be .400", which is a fair number to use for our example since most cams will be close to this, give or take.

DESIGN GEOMETRY 202

Only HALF of this cam lift is relevant because intersecting the "exact" MID-LIFT point is all we need to do; the accuracy for the second half of cam lift is obviously going to be automatic. This means you've only got .200" of movement to dial in on. Any error in this first .200" will be added to the second half cam lift, as explained in DESIGN GEOMETRY 105 (p.37).

If you're pushrod is too short by .050" on the first half, your 90 degree arc occurs .050" LATE, which adds .050" MORE to the second half of your rocker motion. This DELAYS the CAM Velocity to a higher valve lift; or another way to look at it, is that it SLOWS the cam velocity in the first half of its lift, speeding it up more towards higher valve lift. If you are too long by .050", then the opposite occurs, and it leaves the lower lifts faster, and slows more towards full lift. In both examples, you will have nearly the same exact net valve lift, so you won't have a clue how much your cam has changed at the valve unless you map this at the valve with a Digital Trace or old fashioned (slower) graph paper, dial indicator and degree wheel.

The key to this perspective is dividing your error into HALF cam lift. If your 7/16" adjusting screw is off by just ONE TURN for our .400" cam lift example, then you are in error by 25%! However, the "error" is NOT linear in real terms of excessive in and out motion. It's worse.

If you go back to DESIGN GEOMETRY 105, you'll be reminded that the .400" cam lift was shown with a pushrod that was .100" too long (double our example here). If you take another look at DESIGN GEOMETRY 104 (p.36), you'll see the radial motion (in and out) shown was only .024". But as soon as we turned the adjusting screw .100", which is half of the .200" MID-LIFT motion, the radial motion more than DOUBLED, to .054", making an error of 125%. This works out to several extra degrees of crank rotation to move the rocker to the same valve lift. Regardless what valve lift you measure this at, it dilutes throughout the valve lift range, lessening your cam's duration and velocity by a significant amount at all valve lift points.

As noted elsewhere, what you end up doing if you're serious about "trial and error" dyno testing, is putting a bigger (meaner) cam design in until your cylinder heads get what they need, shown in dyno results. But you've over compensated with the cam to restore valve motion lost from excessive over-arcing geometry. However, you're not seeing the lost horsepower from the additional drag coefficients and valve spring surge from the increased harmonics reacting to the excessive in and out motion. So you're still missing extra power, not to mention longevity of all related components that may break, which otherwise wouldn't.

You need to think MICRO in using your adjusting screw, and ordering your pushrods. Ordering pushrods to "close enough" averages from what's lying around on some distributor's shelves is not going to cut it. You need to order EXACTLY what you need, to within .005" or less of whatever your measurement requires for placing the adjusting screw's cup (or ball) axis at precisely MID-LIFT.

With STUD mount rockers, you have no choice. You only have "Installed Geometry" because the CUP is locked in place in the rocker body. Except for the few odd-ball stud mount rockers created with an actual adjusting screw, you have to tolerate the over-arcing inherent with the STUD models made which have this too high pushrod cup. If you find one of these odd-ball adjusting screw stud mount rockers, then you can make your design geometry exactly what you need rotating the adjuster up and down. But be careful of the *angle* the adjusting screw was installed at. Running it down may make things worse. See p.46.

DESIGN GEOMETRY 201

You must know what the adjusting screw's exact position needs to be in the rocker body to precisely create the 90º relationship at MID-LIFT. Then turn it down a quarter turn to allow for valve lash; then "snug" the lock nut without overdoing it. Then use a precise, adjustable pushrod capable of full valve spring pressure without flexing and adjust it to take up the clearance with only a finger tight feel. After removing the rocker, and without disturbing your setting on the pushrod that ideally will have a lock nut, measure the pushrod's exact length and allow for the style of ends it has when you order from your favorite pushrod supplier. If the ends have little flats, you won't be measuring the OAL (overall length) accurately, and it is the OAL terminology that everyone uses for the length reference. Your pushrod supplier should be precise enough to compensate exactly what they need for making your pushrods with oil delivery holes, or any machining flats from the manufacturing process of solid tip pushrods. Most will have oil delivery, so their OAL will measure less with calipers while providing you a few thousandths more in reality for their full radius tip – if it was designed for one. This is how precise you need to be.

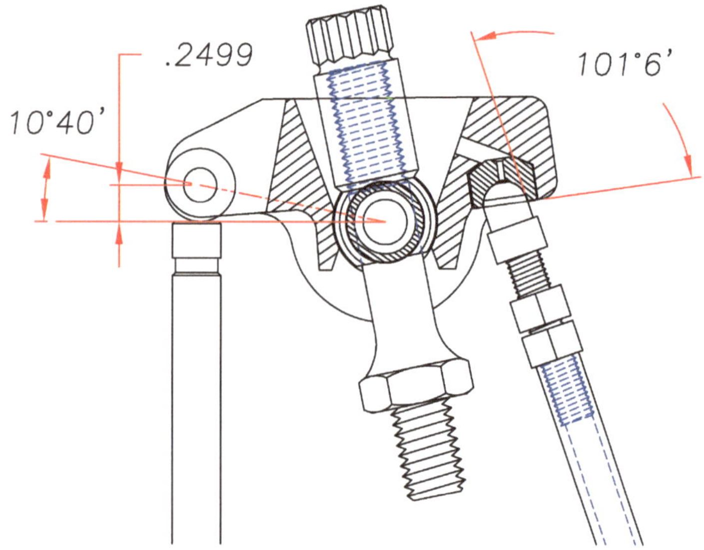

This PVS MID-LIFT rocker sits in the CLOSED valve position. The Installed Geometry is set for .250"; exactly half of the .500" NET VL shown on the complete drawing.

The roller tip indicates a rotation angle of 10º 40' to reach its 90º relationship with the valve, while the pushrod is indicating 11º 8' to reach 90º with the cup's axis.

Unfortunately, most rockers are designed where this cup is already 90º in the CLOSED position, so at full lift it is precariously acute.

The extra 0º 28' pushrod angle depicted by the 101º 06' over the 10º 40' for the valve side is due in part to a lean "inward" the pushrod makes at closed valve. The real in-and-out motion for the .327" CAM lift required for this .500" NET Valve Lift is roughly .016", for an 8.200" long pushrod. The pushrod's actual angular deviation would be 0º 07' (zero degrees, seven minutes). These shifting dynamics are where "some" of the ratio gets lost between "net" and "theoretical" ratios (paper specs), as well as linked to net compression. This is why a "correction factor" in the rocker arm is required to compensate for these losses. Before MID-LIFT came along, no one used a correction factor and net valve lift losses of .05 or more were common from manufacturer's "advertised" ratio; and many still are. The most important point is that having anything LESS than such angular precision for both DESIGN and INSTALLED Geometry means the RATIO will deviate *more* than it needs to as it tracks from Closed Valve to Open Valve.

So, once you know where you want your adjusting screw's cup/ball axis to be – what do you do if your brand rocker arms need their adjusting screw turned down (as most – or all others do)? Well, you can "fix" most of the popular "non-MID-LIFT" rockers to reduce their error; although getting them dead on may not be possible. At least now you can significantly improve them since you finally know how important this is. So what do you look for?

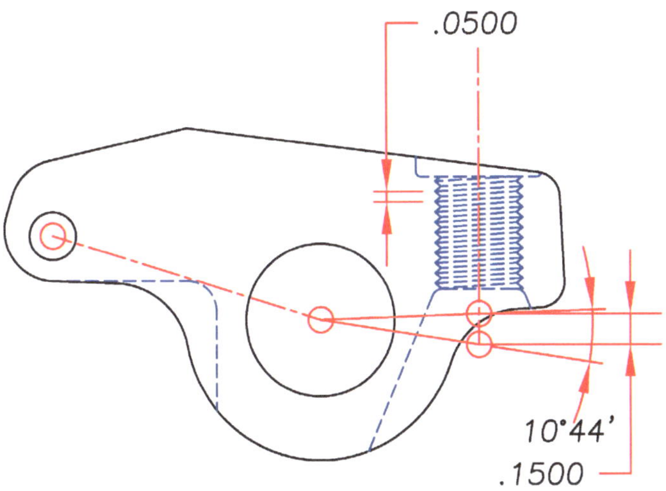

If the adjusting screw is too high in the rocker body, you can rotate it down two or three threads, even though most manufacturers will tell you this may be pushing it, and no doubt argue about keeping any warranty if you break something after telling them this. The reason is they know if their screws are pushed out of the rocker body's support by two or three extra threads, that it will be weakened. This is because they've broke plenty. If the forces are in line with the adjuster, then it doesn't make much difference, unless you go crazy, like five or six threads out. The above shows THREE (3) threads rotation for .150", and you'll notice that it provided a 10 and 3/4 degree increase (44 minutes).

In the view to the right the original adjusting screw length is shown as .796". The second axis datum is the extended length of .150". Its axis LENGTH is .805", for a deviation of only .009" between them. The farther the adjuster is turned down (out), the more this deviation will increase. The math works out to a net ratio loss of 0.02 (two hundredths). With our .400" lift CAM that works out to only.008" net valve lift loss. But because the adjusting screw angle in the upper setting is 90º the deviation will be minimal. What's more important is having the correct design geometry angle, even if it takes turning the adjuster out three (3) full turns to do it.

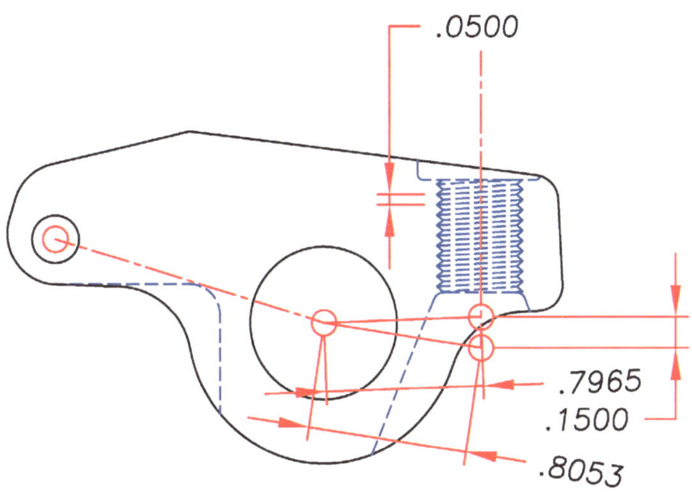

DESIGN GEOMETRY 202

To fully understand what is happening with these and the following illustrations keep in mind that the vertical **RED** dashed line going up the adjusting screw's center is also the angle of our pushrod (not shown). So when we extend the adjusting screw this much, the "deviation" in length from the rocker's shaft won't be too much. That *won't be the case* with rockers that have an acute angle deviation to the pushrod, as most do. That illustration is shown on the last pages, with their effect for modifying.

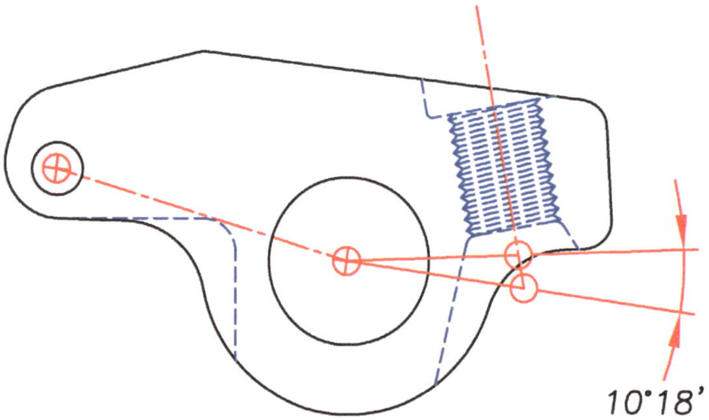

The illustration to the left takes our same rocker from above and lays the adjusting screw axis over 10° to simulate the approximate angle of other systems, which varies depending on the model and engine application.

Rocker systems have traditionally been designed in the "closed valve" position, which had the adjusting screw angle set to the pushrod in this state; so when the rocker rotates to full lift, the majority of its 26° or more rotation is applied to the adjusting screw with the maximum force of full lift spring pressures applied more to the side of the adjusting screw CUP instead of the adjusting screw axis. Oil delivery through it is also impeded.

In the illustration to the right, the dimensions are shown for the rocker above, with the .150" extended CUP axis, same as the preceding page illustrations (DG 202).

The .796" dimension above it is before extending the adjusting screw down three (3) turns.

The .817" dimension illustrates how the length from the shaft of the extended CUP dimension has increased .020", which reduces the RATIO by 0.05, On our .400" cam lift example we would lose .020" net valve lift. Also notice the measurement is skewed to the direct line length, meaning it too is off roughly .012". The greater the adjusting screw misaligns with the pushrod, the more difficult "tweaking" will be. (See #202)

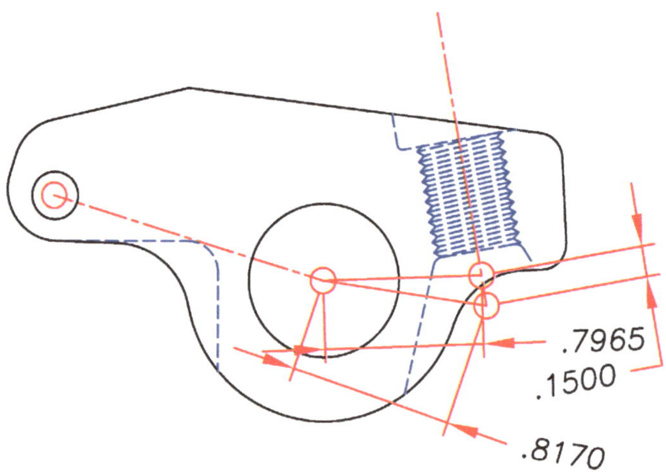

DESIGN GEOMETRY 203

As deterring as making this change may sound, the KEY is to correct the geometry with the pushrod, and accept the other shortcomings with some re-engineering to beef up the weak points. If the "brand" of rocker you're using requires this extended adjusting screw to capture Design Geometry for your engine to duplicate the pushrod to valve angle – as it should – then some of the weakness coming from extending the adjusting screw like this can be fixed by a new design adjuster that is supported beneath the rocker. In ALL CASES, if your "name brand" system suffers the above, **extend with caution**.

Naturally, it would be good if you could take what you now know and choose a system that has these angles already, but be prepared to get the run-around while looking.

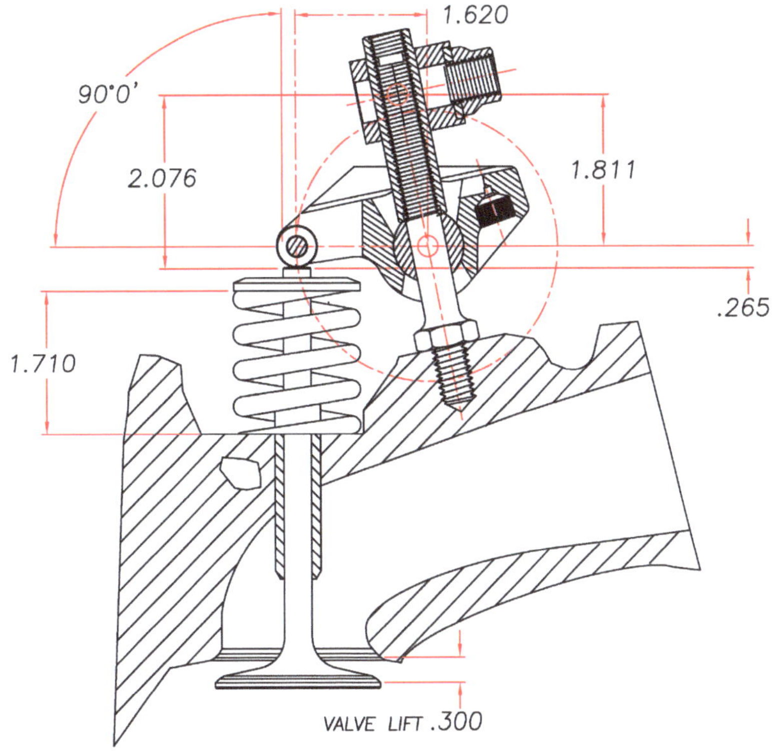

ROCKER MOUNTING

This cutaway illustration depicts a BB Chevy PRO-STUD™ rocker at Mid-Lift. The valve is open half of its .600" total, and the valve spring is almost 1.710", or .300" from its 2.010 installed height. NOTE the trunnion axis relative to the roller tip axis, which is in-line horizontally as depicted by the **RED** dashed line. Other dimensions (2.076" and 1.811") relate to the Stud-Bridge™ mounting height. To the right, the **.265"** gives the depth of the valve tip below the trunnion axis, which is HALF the ROLLER DIAMETER. Since half the Valve Lift is .300", subtracting .265" means the valve tip will be **.035"** ABOVE the trunnion axis when CLOSED; which is the best way to measure and set your installed geometry.

The rocker stud leans into the valve most OHV engines, and as the above shows, it is easy to see that the higher the rocker arm to accommodate longer valves, the shorter the distance between the trunnion and the valve, forcing the roller to run on the outer half of the valve stem tip.

It is important to state again that no matter where the roller tip sits on the valve, what is important is the angle of pressure, which needs to be at this 90 degree angle at MID-LIFT. As long as the roller is somewhere in the middle of the valve tip it doesn't matter if it is .050" bias to one side of the valve center-line or the other. Valve guides are usually over 2.50" long. The valve stem is typically .342" in diameter. This provides a valve stem support of more than 7:1 for the loads being applied to push down on the valve. The valve does not care about the point of pressure being .050" on one side of the valve centerline or the other. But it does care about the "angles of load." If the roller is rolling across the valve tip instead of pushing straight down on it for the majority of its arcing motion *and with the least amount of that motion*, then the valve stem and guide will see side loads and increased harmonics that go with it, even though in theory there is a "roller" tip. Why? Because the roller does NOT roll anyway. The roller is only a pivot point of rotation.

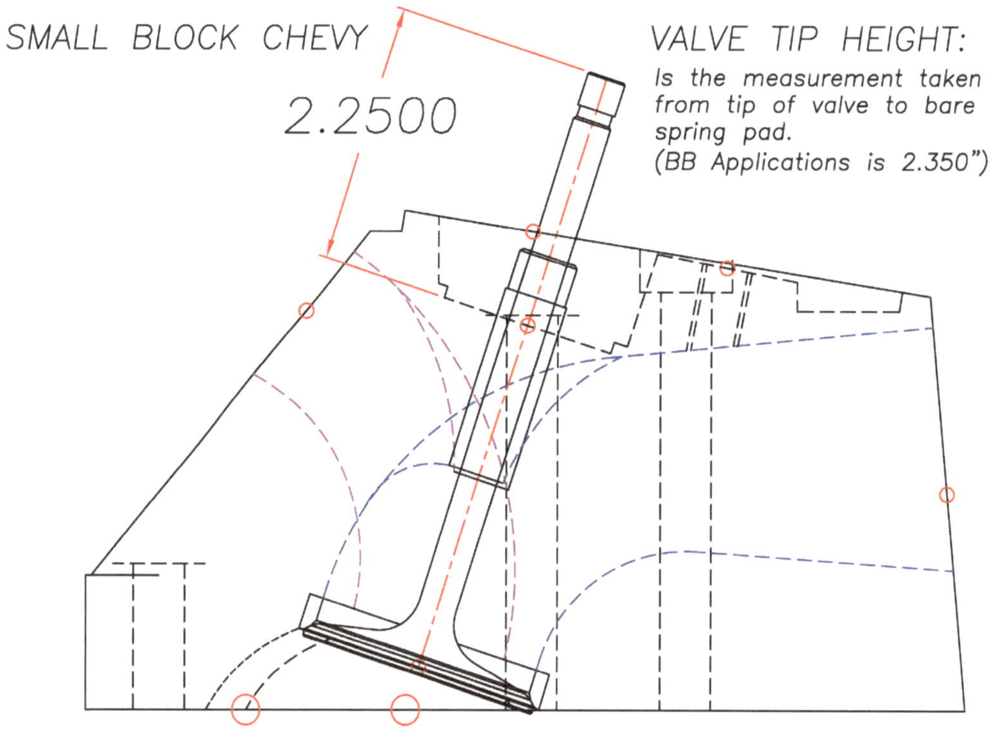

VALVE TIP HEIGHT

VALVE TIP HEIGHT ("VTH") is a critical standard with rocker arm lengths. The randomly chosen valve lengths that are often selected for only attaining spring clearances, are often over done. When an engine builder mistakenly assumes "middle of the valve" is correct geometry, they are often making a too long rocker over-arc. Setting its angle correctly will often put the pattern across the valve's center-line. MEI rockers are made more accurately to this standard, where such problems don't exist.

For too long, ".100" longer," and ".200" longer" than stock valves have been the normal mode of thinking for engine builders trying to gain extra valve spring clearance. But with the myriads of head manufacturers, models, and changes, it is crazy to fit rocker arms to all these combinations. Since head preparation was the main business I was in for 15 years while developing MID-LIFT Geometry, this was a natural question – and answer for me. Standardize to the valve spring pads. These used to be sacred ground anyways.

Back in the "old days" of "292" Chevy head castings, it was a very expensive mistake to cut through the spring pads for extra spring clearance. Because everyone has become so used to just ordering longer valves, few people really cut the valve spring pads anyway, and it is a dimension that remains relatively constant from head model to model of each manufacturer. But that doesn't mean the same intended design from two different manufacturers will be the same, and probably isn't. Chances are that the same part numbers you bought this year will have the same specs next year if you're using the same aftermarket cylinder head manufacturer. Only OEM part numbers are consistent (usually).

VALVE TIP HEIGHT

By knowing what the various valve spring retainers and valve spring clearances are for the optimum valve tip length needed, once the spring cups, shims and other related components are stacked up, I had the informed insight to decide on a Valve Tip Height "standard" for which all rocker designs would be based upon. The above is for a full competition small block engine up to .800" Valve lift (which is beyond the usable valve lift needed). Adding .100" (2.350") for Big Blocks, can then get close to 1.000" (one inch) Valve Lift, while keeping coil bind specs for your springs to a .100" clearance, the absolute minimum I ever used, in spite of some who cite only .050". **That's too tight**. Engines in the 1" valve lift range require a fully dedicated rocker design specific to their exact engine, not a cookie cutter design! For small block engines that will see valve lifts up to .750", **2.250"** is what the designed **VTH** is. For big block engines, running valve lifts of .850", this is increased by another .100" to **2.350"**, as noted above. Needless to say, on the crazy valve lifts of .900"+, another .100" is used. But these two main standards have always worked since 1980, when I locked these in for heads and engines we prepared.

> NOTE: The above is for high valve lift, professional racing; MANY small block applications are fine with a 2.150" VTH, to avoid excessive "shimming" beneath the valve spring cup. The 2.150" VTH may better be suited for installed spring heights of under 2.00", where valve lifts will be .650" or less, OR... valve spring pressures are under 175 lbs, as with hydraulic roller applications. In either of these examples installed spring heights over 2.00" are not likely. But for a standard which precision geometry rocker arms are designed to, the above 2.250" is chosen to centrally place the roller pattern upon the valve, and easily accommodates shifting this pattern .020" across the valve's centerline for this short VTH, or behind the valve centerline for an even higher VHT, such as 2.350".

Valves are made to specific lengths that are derived from the spring pads of the heads they are going to fit in. There is no mass production of this, unless the measurements are closely controlled on all components. That means valve seat depths from the deck, spring pad heights from the deck, and spring pad heights from the deck. These "gauge" points are represented by the red circles in the previous page illustration. On the small block Chevrolet, the red circle at the lower left corner of the above illustration is the DOWEL point, which is where all critical dimensions originate.

The other critical dimension that is available and used two ways, is the **valve tip length**. Ford uses **.310"** while Chevrolet used. 250". Many aftermarket valves come standard with the Chevrolet tip length, unless you specify otherwise. We highly recommend the .300" (.310") length. When ordering your valves, you must be careful about how you give your dimensions. Good valve manufacturers will know how to interpret your needs, but why shoot in the dark. Valve specs refer to "overall lengths" which include dimensions you may want to allow for, or not allow for. If measuring your valve length from the valve SEAT of the HEAD, then you need to ADD the MARGIN width and the BEVEL height on the end of the valve head.

> NOTE: As explained in the next illustration, if using a more common aftermarket valves .250" tip length is preferred, then the 2.250" (Small Block) and 2.350" (Big Block) dimensions can be reduced .050" with adequate clearances in all but the most radical cases. However, retainer selection to get the desired valve tip clearance will have to be chosen carefully.

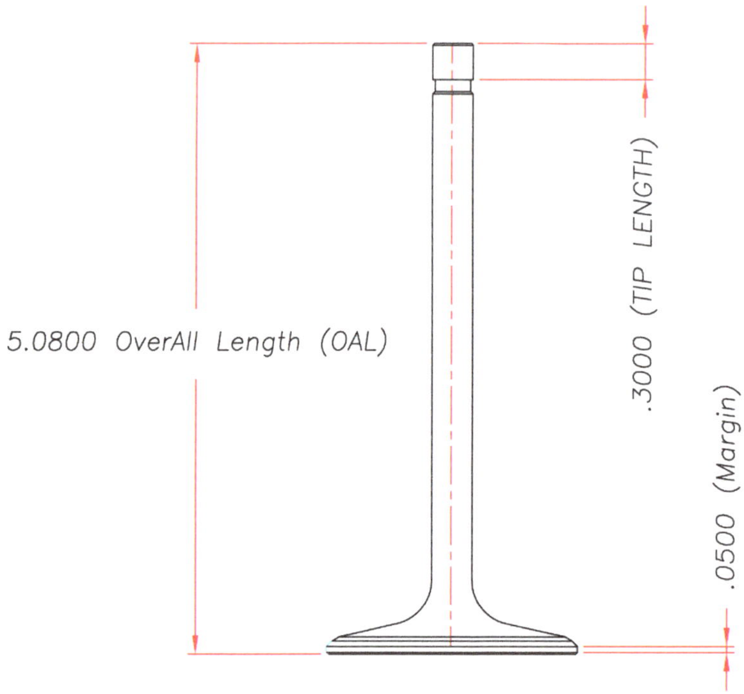

VALVE TIP LENGTH & OVERALL LENGTH

The **Overall Length** shown here allows for dimensions that you might be removing if you measured the spring pads from the valve seat, as the author used to do. In this case, you would add the .050" Margin height, as well as another .020" for the BEVEL edge on the valve's combustion chamber side. The valve MARGIN is best around .050" for the intake, and .080" to .100" for the exhaust valves. IF a LASH CAP is used on your particular assembly, you can allow for that thickness, in lieu of the .300" shown.

Choosing valves is very important for more than just performance; they are critical in establishing how the rocker arm operates; or more specifically: where you make the rocker function.

VALVE TIP HEIGHT needs to be standardized and it can't go by the adaptive terms of ".100" long" if rocker designs are going to work, because this is ambiguous and deceiving. Longer than what? Heads with reworked valve seat locations, or angle milled, welded, etc., move these specs around. There was no standard where a "loose" term like "a hundred long" will be the same from one head to the other. Measuring from the sacred valve SPRING PAD is best, because is a common dimension. So long ago I chose the Ford Valve TIP specification of .310" to establish where the keeper grooves would be. The standard height retainer was also offered in a higher and lower version, so it was chosen with the .300" valve tip length. The crown of the valve tip sticking above the retainer is perfect for the rocker arm's underside clearance. However, if you prefer to use the shorter .250" Chevrolet valve tip length, then just subtract .050" from 2.250" shown (for small block engines) for a 2.200" maximum spring height. For BB Chevy specs with a .250" TIP, simply reduce this from the 2.350" to 2.300" that's suggested in VALVE TIP HEIGHT on the preceding page.

The retainer and valve tip views shown on the next page can still be achieved with these small valve TIP lengths, but you may have to be more selective in choosing your **RETAINERS**. Remember, with custom valves, you can have exactly what you want, but these specs will work for most applications.

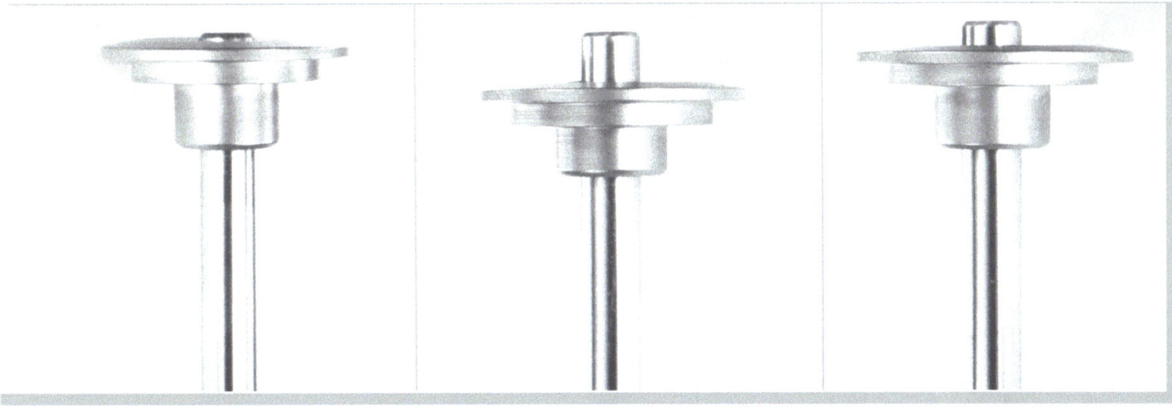

TOO SHALLOW	TOO HIGH	PERFECT

LEFT: Shows a valve tip which is TOO SHALLOW using a STANDARD depth retainer with a .250" Chevrolet Valve Tip. The more shallow retainer (center) can solve this, but crowds the keepers and has limited choices in keeper styles for strength. CENTER: Shows a shallow retainer with a .300" tip.

RIGHT: Is a "**PERFECT"** combination of the STANDARD depth retainer on a .300" valve tip. Here both strength and choices in keeper designs are adequate for various installed spring heights and valve tip rise above the keeper is maintained.

RETAINERS & VALVE TIPS

Any of the above examples have been used successfully, but you must be aware of the pitfalls the first and second will have. Too shallow means the valve spring retainer can crowd the underside of the rocker arm, forcing some creative engine builders to "notch" it (taboo).

The second example, assuming the desired valve spring height was achieved, means the valve tip is too long (high), and is forcing the rocker to go much higher too, in order to establish the correct pivot point height with the valve tip. The IDEAL thing here would be to replace the valves, if budget and time allow. Even longer pushrods will be required to get the correct geometry, beyond what otherwise would be needed with a shorter valve. In the case of the small block Ford and Chevy with the inclined studs this also means the rocker's tip will be shoved further across the centerline to find the correct geometry.

I've noted before that this last consequence doesn't really matter. The top priority is making sure the contact pattern remains totally on top of the valve tip – and providing the geometry that is desired. Often this will have the wear pattern shifted to the outside, or even (but rarely) to the inside of the valve tip centerline. Don't worry about this, just make sure the "angles" of motion are your priority, and that you have the 90° relationship that you need at *exact* mid-lift.

But why go this way if you don't need? Either fix this now, or you will compound the costs to fix it later when you'll be buying another set of pushrods for the more appropriate length valve. And if you chose such a valve tip length for some clearance issue with the rocker arm's underside, then you've got the wrong rockers! As you go to smaller valve stem diameters the dimensions of the above examples will naturally change, but their formulas won't.

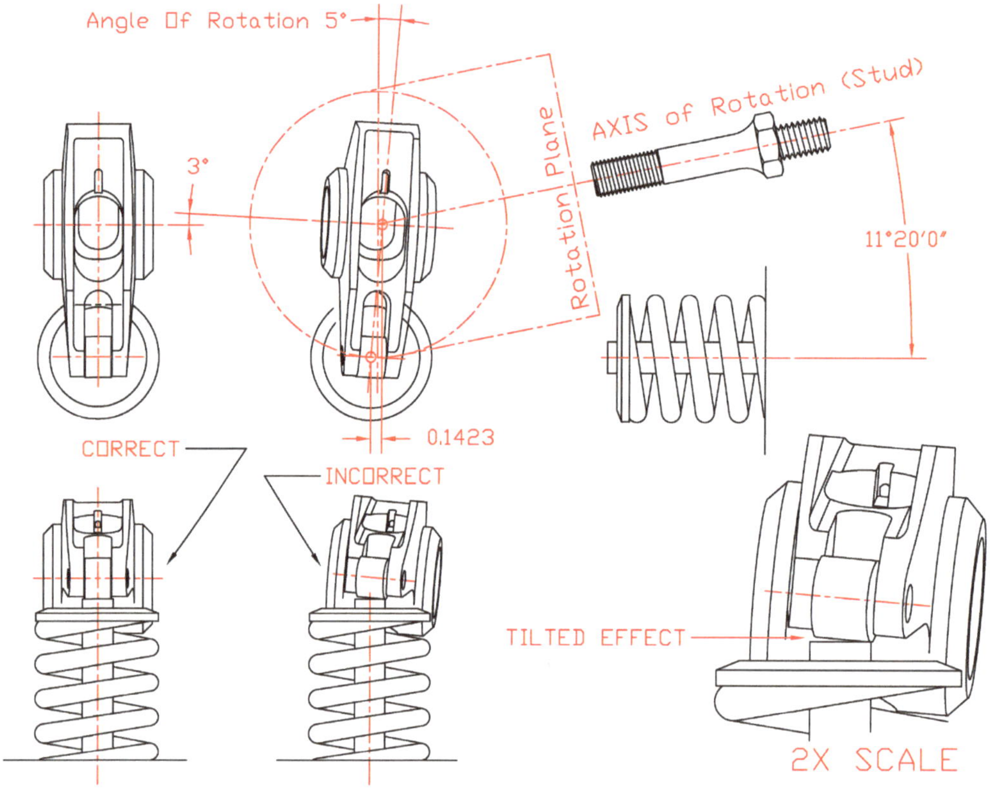

TWISTED ROCKERS

The left side of the Illustration above shows two views. The first shows a rocker with the valve, stud and pushrod cup all in-line. Everything is fine. The second rocker arm has a stud that is .142" offset from the valve, and on this example length, it creates about 5 degrees of rotation. Some head models may have less, some more. But the point is, any misalignment that causes the rocker to be rotated will still show the same symptom of damage from misalignment of the roller upon the valve tip. The valve spring and stud illustrated in the upper right shows how the STUD Axis leans into the valve 11° 20' (11-1/3°).

Whenever a rocker arm is ROTATED on the stud of a head that has IN-LINE valves (as viewed from the end of the head), this is the effect that is created. There's no other way around it. Yet, to this day, countless heads are sold with offset pushrods, and WORSE, offset studs that FORCE the engine builder to "presume" this is for a proper design reason, when in fact it is GROSSLY bad engineering, or bad judgment. Any heads that have studs with a compromised offset from what the pushrod is, induces false trust to the engine builder that a stud mounted rocker will operate properly as long as the three points are connected: PUSHROD, STUD and VALVE. Not so! You will need STUD rockers that have a designed offset in the roller and pushrod cup, to use with such heads. OR... you will need STAND mounted rockers (also known as "shaft" rockers).

If you don't, then just like the illustration above, the roller will be tilting to one side and only applying pressure by its side edge to the valve in a side-loading arc, regardless of what you do to the rocker height. Adjustments to the rocker height will only change the point of valve lift where maximum MISALIGNMENT occurs, but it will occur! Do yourself a favor, buy stand mounted rockers, or buy stud mounted rockers with the offsets on both the roller and pushrod, or better yet: if you are determined to use stud mounted rockers and can have the option from the head manufacturer, **BUY the HEADS with the rocker studs drilled and tapped on the SAME Center-line as the valve spacing**. Leave the pushrod offset entirely in the rocker.

TWISTED ROCKERS

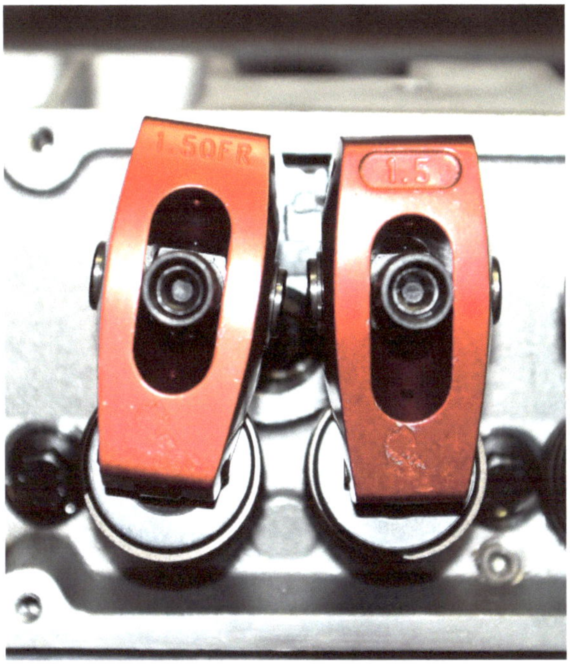

The above is a great example of what NOT to do!

The budget brand rockers shown above and sold under several "private label" trade names even have offset pushrod cups, which should have minimized the above "twist" to aid in a very limited forethought head design. But the choice of parts was to no avail.

In spite of this, these aftermarket heads have such a misaligning STUD placement for the pushrod offset that this TWIST is the BEST result that could have been obtained; and a stud mount rocker with conventional center located pushrod cup would have been impossible. This head needed a STAND mount system specific to these specs.

The best way to use a STUD mounted rocker arm where a pushrod offset for wider ports was required by the cylinder head manufacturer is to have *both* the STUD and VALVE centerlines in the HEAD be the same C-C dimension from intake to exhaust, and cylinder to cylinder. *Then...* use a rocker arm with the pushrod cup offset as needed, to avoid a twisted setup as the example above. This requires the head manufacturer to make an equal stud and valve center-to-center design available, if only as an option; then secondly, a rocker arm manufacturer who provides the needed cup offset.

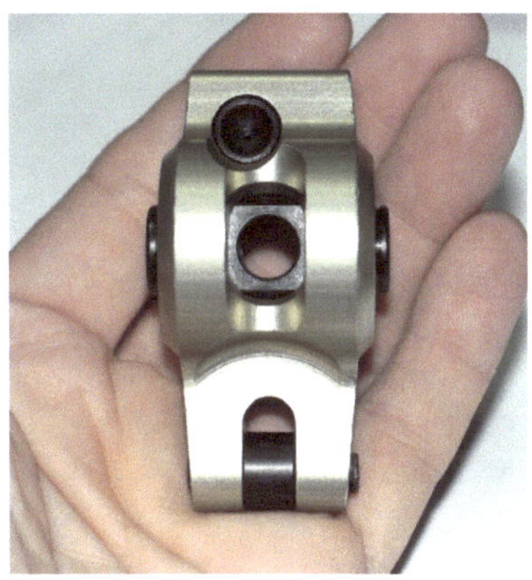

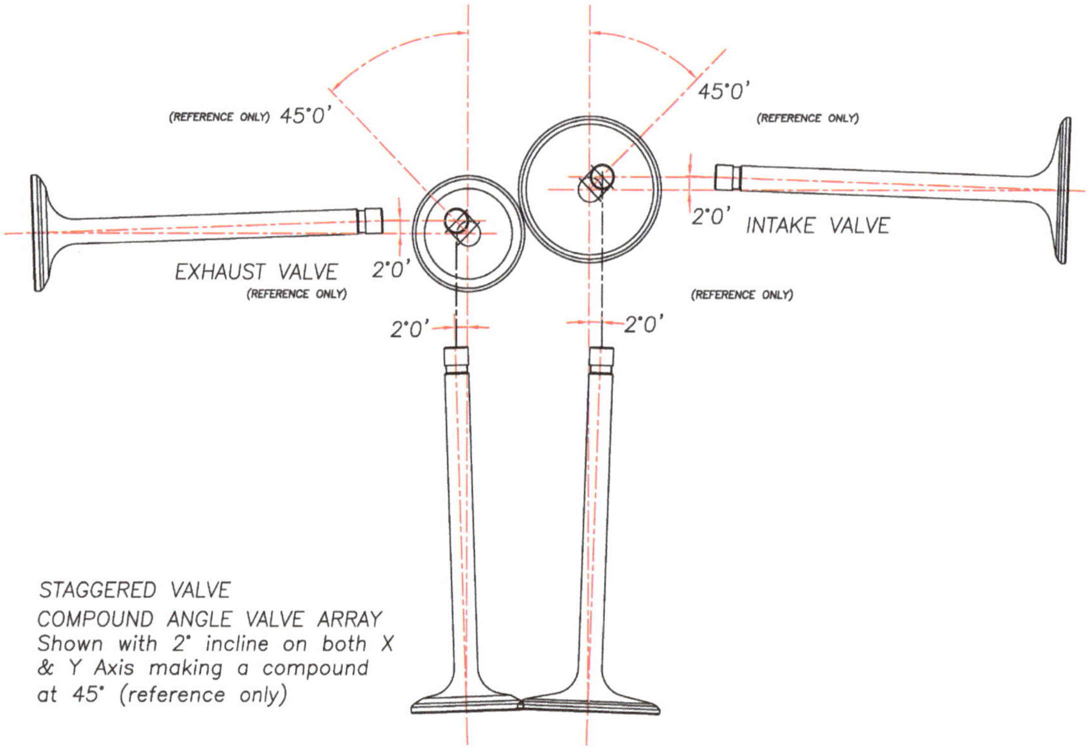

COMPOUND GEOMETRY

COMPOUND GEOMETRY, in cylinder heads refers to the tilting of the valves on more than 1 axis. Here, each valve is tilted on 2 axis` to illustrate the 3rd angle that develops. Each is laid over 2 degrees, on the X-Axis (length of head), and the Y-Axis (width of head), to create a third angle which is "compound" to the preceding two.

Rocker arms must be laid over to accompanying angles that provide the roller sits flat atop the valve stem, while still positioning the push-rod at a compatible angle so the side thrust are minimal. Unfortunately, this fails on most designs. Which is why we change it.

The only time a rocker arm should be twisted on the stud, is when this type of valve array is being used. This is an example of BOTH a "splayed" and "staggered" valve array. Staggered is when the exhaust and intake valves are shifted with each other in distance from the intake manifold (the "Y" axis). Splayed is when the are laid over to the left and right, as viewed "over the fender" of a car (the "X" axis).

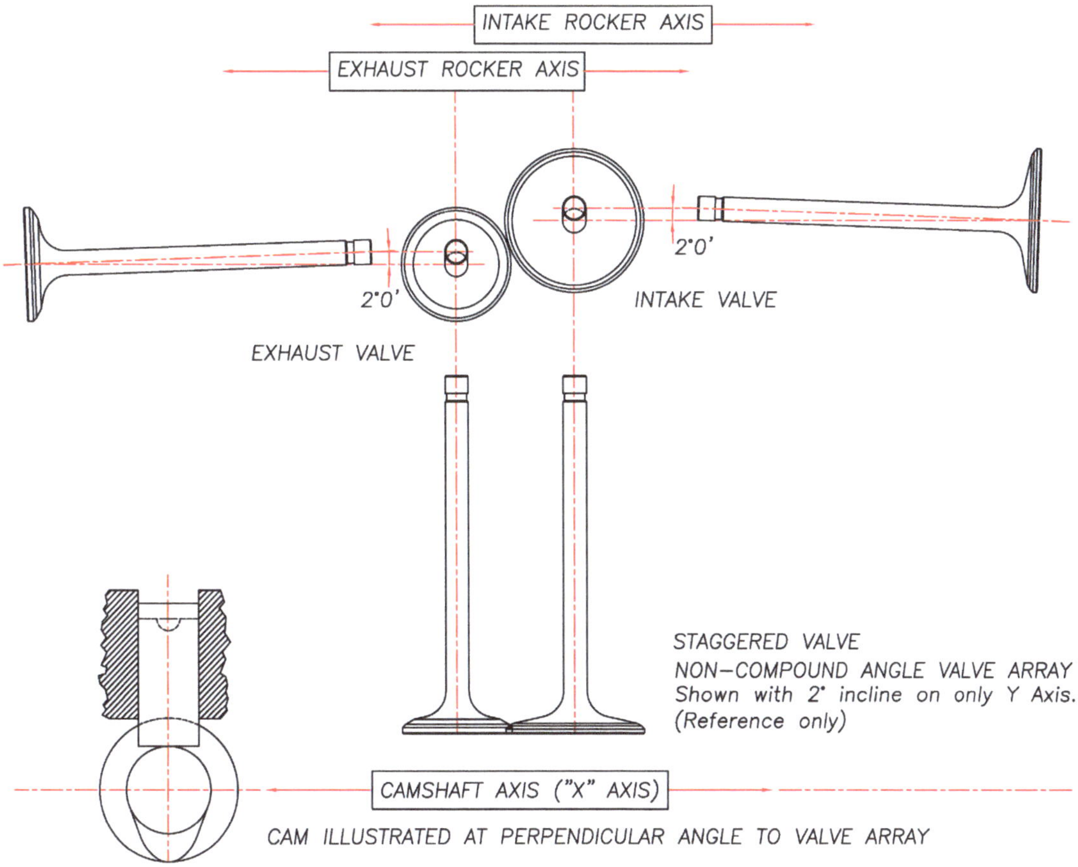

STAGGERED VALVE ARRAY

STAGGERED VALVE ARRAY, in cylinder heads refers to the offsetting of the intake and exhaust valves with each other, when viewed from the end of the cylinder head. Here, each valve is tilted on 1 axis (Y-Axis). Each valve is laid over 2 degrees, on the Y-Axis (Width of head), and the X-Axis (Length of head) is NOT leaned; but is shown 90 degrees to the cam's axis.

The rocker arm's axis must be KEPT PARALLEL to the CAM's AXIS on these type of valve arrays, that are still STUD mounted. More complex shaft systems can find an alternate 3rd angle, but their efficiency is not improved in doing this.

There is never any time a rocker arm should be twisted on the stud, unless the stud has been positioned on a compound geometry to work with the compound geometry of the valve. The above is NOT a compound geometry valve array. In fact, this STAGGERED Valve array falls under the same rules for keeping the rocker's axis parallel to the cam, as the in-line valve array of a small block Chevrolet or Ford.

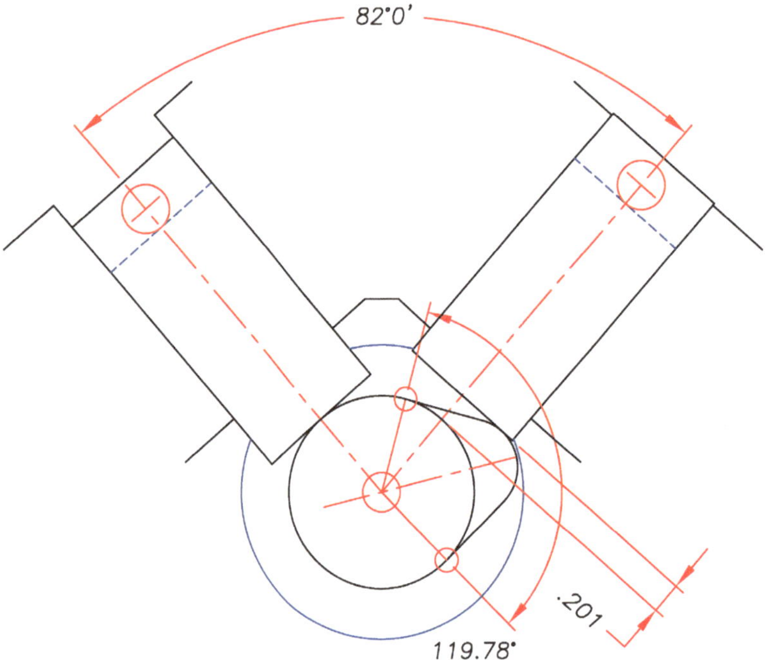

CAM BASICS

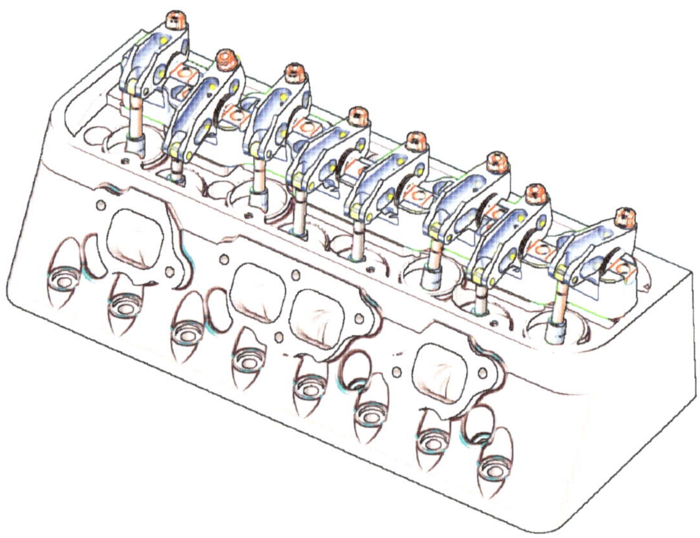

...A little more than just "Basic"

CAM BASICS
INTRODUCTION

CAMS are one of the first and most popular items changed for a sportsman level high performance engine, perhaps second only to rocker arms; but with professional level racing, they are an integral part of the entire engine package that is literally the heart of the engine by which the success of everything else is determined. Numerous companies sell camshafts, and more and more they've been sold as "private label" by non-manufacturing companies – which are usually warehouses developing their own brand – to which more and more an endless supply of endless choices can confuse buyers more than supply a needed product that is at the top of its game.

Engine builders are forced to "trust" not only the knowledge of the salesperson at the cam company, but also that they understand the engine builder's need. That means, what competition the engine will be used for, what heads are on it, what compression it has, what manifold and fuel system is being used, and of course the cam salesman must have a crystal ball to the capabilities of the buyer on the phone they're trying to convince which cam they need.

The cam is the "heart" of the engine. The cylinder heads ARE the engine, and everything else is just mechanics for making the heads work. But the cam is responsible for the synchronizing of everything the heads need to move air. When to begin opening the valves; how far to open them – and for how long to keep them open. Rocker Geometry is a critical element for delivering this information accurately, fully and efficiently. The Tech documents "Installed Geometry" and "Design Geometry" illustrate how critical this is, but you still have to grasp what your choices in camshafts are, and why, if you want to avoid the pitfalls of "trust" for someone who can only – at best – get you in the ballpark.

Among the information here that has been revised from my original publishing of CAM BASICS are some general rules for cam recommendations, as well as an overview of what type of cam to use – without getting oversold by a cam company on something that you not only don't need – but can't fully use. If you don't know that less is more, then you really have no way to measure what you *didn't* get anyway.

This chapter is for engine builders who want to be informed on options for cam selection. As with everything in building competition engines there are exceptions to most choices, *except rocker arm geometry*. The recommendations here are a good foundation to begin with to avoid many problems, while arming the reader with tools to avoid confusion about what's really important on the most critical component of the engine that can make or break its performance success. *~jM*

Note: This chapter may contain some repetition of points from other chapters. For continuity most of these have been edited wherever possible, but some have been left intact for emphasis where relevant.

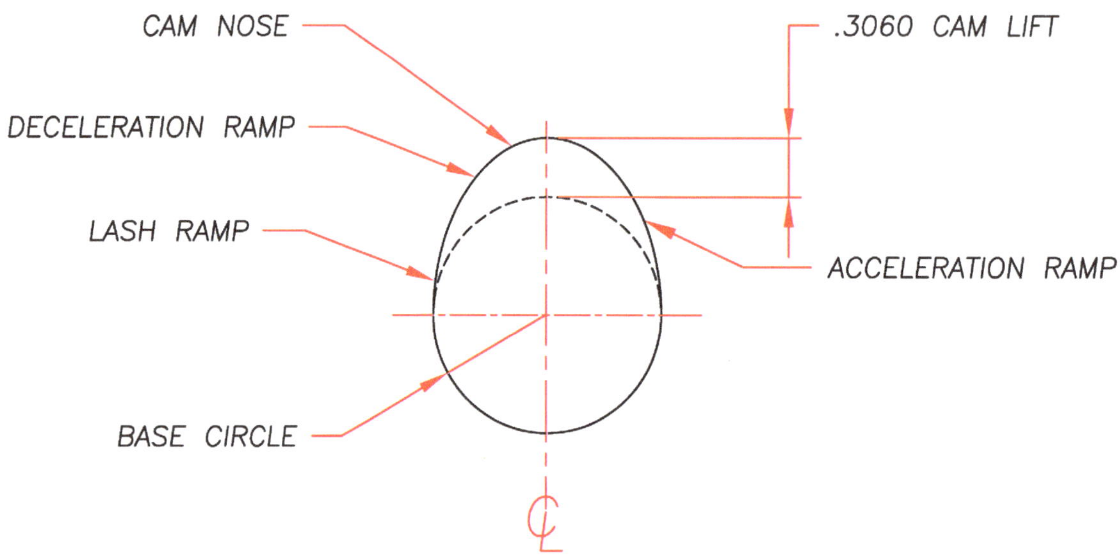

CAM LOBE BASICS

CAM LOBE is shown with four major components: **BASE CIRCLE, LASH RAMP, ACCELERATION RAMP & CAM NOSE. CAM LIFT** is shown as .306".

The CAM RAMPS are "acceleration" and "deceleration" with specific and precisely calculated profiles aimed at "controlled" opening and closing of the valve, in a quick, long and smooth manner. The above shows the opening side (on the right) ramp as "acceleration," and the opposite side as "deceleration." The fact is, *each side* of the cam's flanks are **both** and acceleration *and* deceleration.

Not to be confused with "opening" and "closing," the cam's job involves a controlled rate of accelerating the tappet to its ultimate lift, but slowing down that acceleration *before* it gets there, which is when the deceleration comes in. The closing side ramp of most top competition cams begins the closing process quickly; then, like a blossoming parachute that opens at the last minute, sets the tappet (valve) down as gently as possible.

Sounds easy, but it isn't. Mass, inertia, harmonics and physical constraints imposed by the cam's bearing journal diameter and cam follower diameter, dictate limits that challenge good cam designers. The **cam lobe centerline** of old designs had identical and opposite (**symmetrical**) profiles on either side of this line; where newer, high tech profiles displace separate opening and closing profiles (**asymmetrical**). Such designs usually open the valve quickly and set the valve down more gently.

However, the effectiveness of these ramp design intentions at the cam are diluted at the valve by improper rocker arm geometry that is either over-arcing on the pushrod side, or the valve tip side, or both. See: ***Design Geometry*** (p.29).

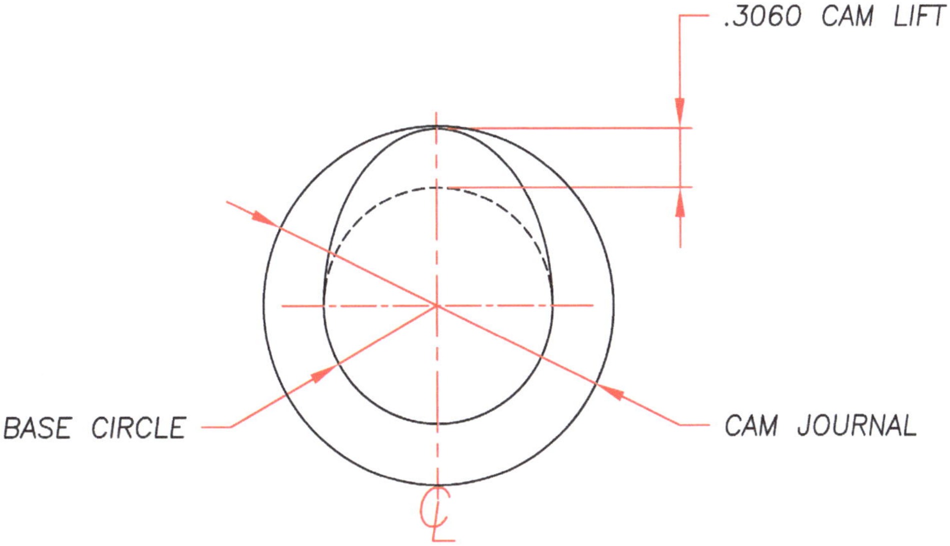

CAM LOBE LIMITS

CAM LIFT is shown with the main limiting element of design, the CAM JOURNAL.

All modern American OHV engines with a flat tappet cam follower (Lifter) maximize their cam lobe lift through keeping the NOSE up to the maximum height of the cam bearing's journal diameter, minus a few thousandths.

Actually, this is true for roller tappet cam followers, too. However, their need for this to achieve maximum velocity at any given ramp angle is not dependant on the nose of the cam being at the bearing ID limits, as flat tappet cam followers require.

The roller tappet cam lobes are shoved to these similar bearing ID limits for longevity and flexibility, since their base circles will be larger to accommodate a given velocity. In other words, the higher the nose (for either design), the greater the base circle can be for a given velocity. The difference between the two concepts (flat tappet vs. roller), as mentioned elsewhere, is the rotating angle of attack a roller wheel imposes on the cam lobe (as illustrated on pp.85–86).

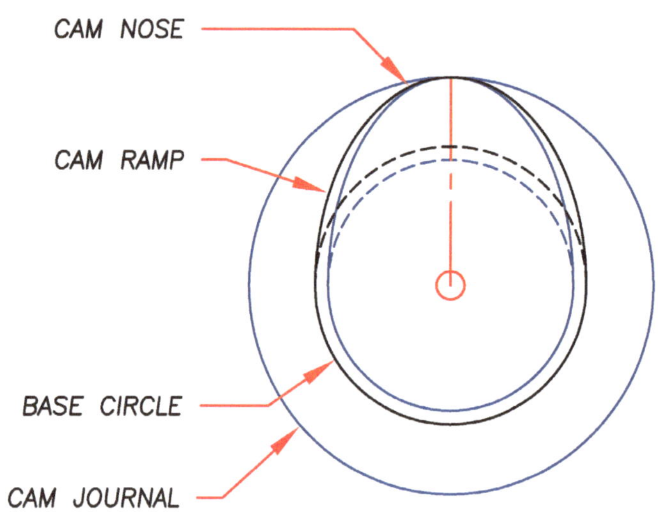

CAM LOBE LIFT

CAM LIFT is shown here with the cam's BASE CIRCLE being REDUCED to create the new cam lobe (inside it) by simply taking metal away from everywhere but the nose.

CAM LOBE is encircled in BLUE which represents the CAM BEARING'S inside diameter. This is the cam lobe's ultimate limitation. TAPPET DIAMETER (not shown) represents the cam lobe's greatest acceleration rate limitation (except for roller tappets, which have no limitation).

All cams are ground with their nose taking maximum height from the inside diameter established by the cam bearing journals. Whenever a cam is ground for an increased lift, the BASE CIRCLE is where the material is taken from, so that the nose is always at the maximum INSIDE Diameter of the Cam Bearing Journal. This creates problems on bigger cams, especially those which will use a FLAT TAPPET (either hydraulic or solid), because the tappet diameter directly limits the degree of acceleration the RAMPS can be ground to. Or more simply stated, the INCLINE of them to raise the tappet quickly coming off the base circle can only be ground to the outer diameter of the tappet. ROLLER tappets do NOT have this problem.

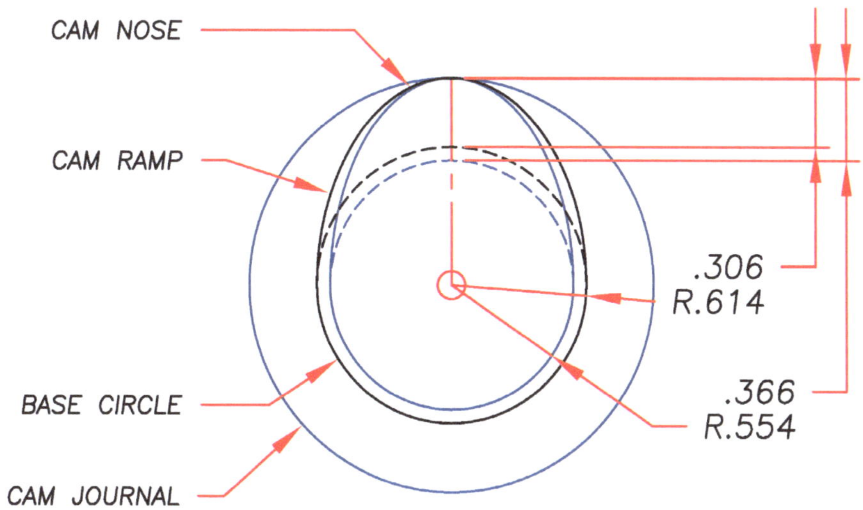

CAM LOBE LIFT

CAM LIFT is shown here as being increased from .306" to .366" with the cam's BASE CIRCLE being REDUCED .060" to create the new cam lobe profile. The base circle dimensions in our example above are represented by the "R" dimensions of the .614" "radius" and .554" radius.

Since all CAM LIFT is derived from REDUCING the BASE CIRCLE and blending the new ramps into the existing NOSE to keep this dimension as high as possible, this of course means the tappet will sit lower in the block. But of course all the tappets see in operation is the degree of rise that the lobe ascends from the BASE CIRCLE.

Engine builders who haven't dialed in on the various dynamics to setting rocker geometry forget that changing to a bigger lift camshaft means the tappets sit lower in the block, so the pushrods sit lower too, and assuming the geometry was close to correct before the cam change, now the pushrods need to be lengthened by whatever the base circle was reduced. Otherwise, your new cam just changed TWICE. Once when you put in the new cam, and again when the rocker pivot points moved because you didn't change the pushrods to compensate. Maybe better, maybe worse. Who knows? You don't, unless you establish the geometry to the same standard for both!

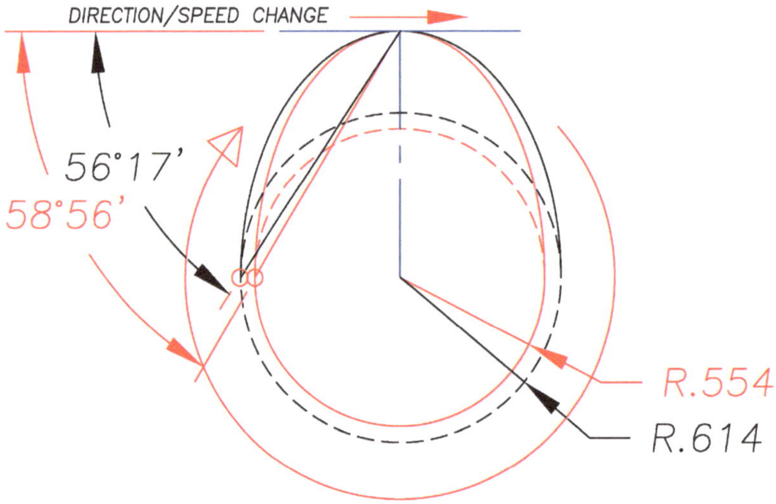

CAM LOBE ACCELERATION

CAM LOBE ACCELERATION has more limits as the base circle is reduced; not so much in reaching the desired acceleration velocities, but in slowing them down. Tappet diameter also works with the base circle in setting cam velocity limits, since this dimension establishes the limits of the cam lobe's reach without running of the edge of the tappet's outside diameter. Greater lift and velocity requires more cam lobe swipe along the bottom of the flat tappet's surface; which isn't really flat, but looks like it is.

Although the above is a simple perspective to a process of cam ramp design that requires extensive computer generated geometry to blend the precision ground curves for high speed control of tappet mass and valve-train inertia, the above example is still relevant.

The "**T**" formed by vertical centerline and the horizontal line at the top of the cam lobe represents the level point where the inertia and speed of the tappet must reverse directions. Unfortunately, this is not reality. It is the ideal. Slowing down must be predicted well in advance, and this is where the *science* and *art* of good cams happens.

The two angles illustrate the how reducing the cam's base circle increases the vertical attack on the **T**'s intersection point. Even though in reality these ramps are not a straight line, the effect of their symptoms from a smaller base circle is still the same.

As noted elsewhere, these limits pertain to flat tappet profiles since roller tappet geometry lives by different rules. But unless there is a need for acceleration rates in a roller that makes the extra costs and working forces worthwhile, the simpler the better. To that end, the basic flat tappet mechanical cam is as simple and efficient as possible, and the best bang for the buck. With proper minimal care it will last for many years.

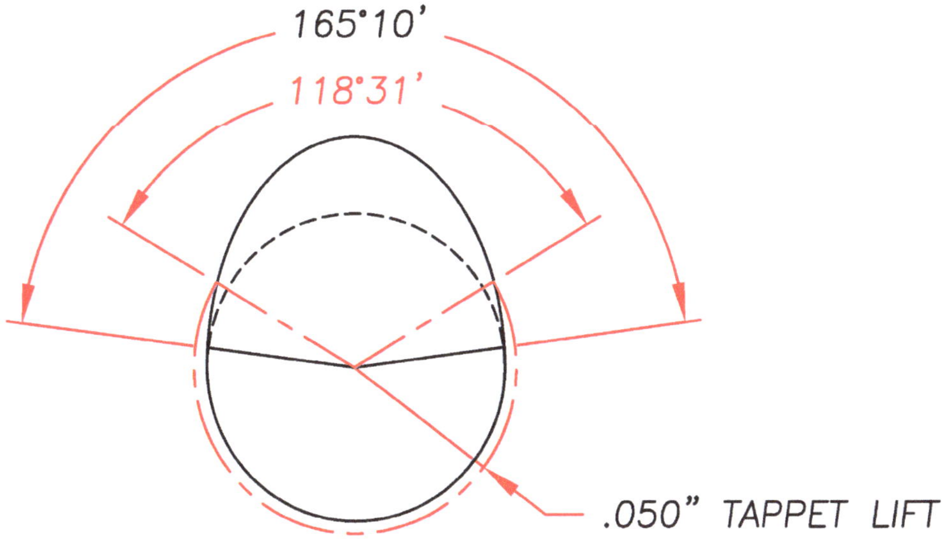

CAMS 101-A

CAMS and rocker arms are integral to the success of the valve-train; especially with Mid-Lift geometry. The 118°31' depicts the DURATION in CAM DEGREES @ .050" tappet lift, which is HALF of your CRANK degrees, also what the CAM CARD would reference to. Accordingly, this example would be **237°** Duration @ .050" Tappet Lift.

The dashed **RED** semi-circle beneath the cam's **BASE CIRCLE** illustrates the tangent points where .050" lift measuring points stop and start on the lobe.

The **BLACK** lines extending from the 165° are the **GROSS** Duration. Doubled, this is 330°. The *gross duration* is meaningless. It is inclusive of the lash ramp, which is a *moving target* from cam to cam.

Lash ramps – which really don't even exist per se, are only the stepping stone from ZERO lift up to when the velocity of lift is deemed usable for achieving the desired *area-under-the-curve*. (The usable breathing area of valve lift for a given head/engine/RPM combination.) The .050" tappet lift is far enough into the tappet lifting cycle that this point has already occurred, regardless of the cam profile; which is why it has become the standard for reference of one cam's true duration energy versus another.

CAM LIFT is not shown, but is the same .306" of our prior illustrations. On a rocker arm with a 1.50:1 NET ratio, that would provide a modest .459" Gross Valve Lift.

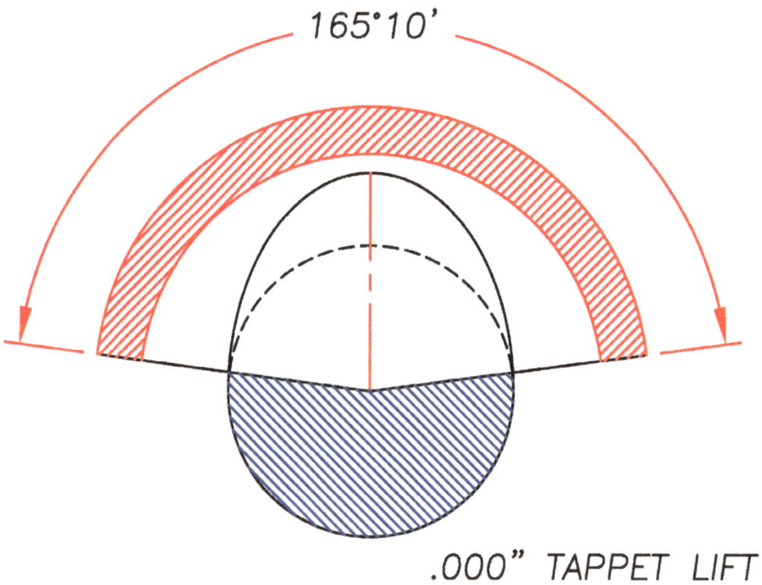

CAMS 101-B

DURATION is shown as 165°10'. To compare with the terminology and specs that would be used in selling this cam, or checking it on the engine, this figure would be double, or **330°**. But this is the measurement perspective the cam is engineered from. The **CAM LOBE** is represented in **WHITE** lines, with the white dashed line illustrating the base circle in contrast to the cam lobe, if it was to continue around, which on a billet cam it does (in most cases) between the lobes.

DURATION of the **FULL CAM** is shown here as 165°10' for a reference measuring lift of .000". This is **GROSS DURATION**, and is more aptly shown on cam cards relative to HYDRAULIC cam grinds.

CAM DURATION is shown here in TWO perspectives, essentially identical to 101-B, however here we are wrapping up this duration sequence by shading in the active cam lobe area. Again, note the "TAPPET LIFT" is ".000"" on our illustration, for GROSS DURATION, as a hydraulic grind cam might indicate.

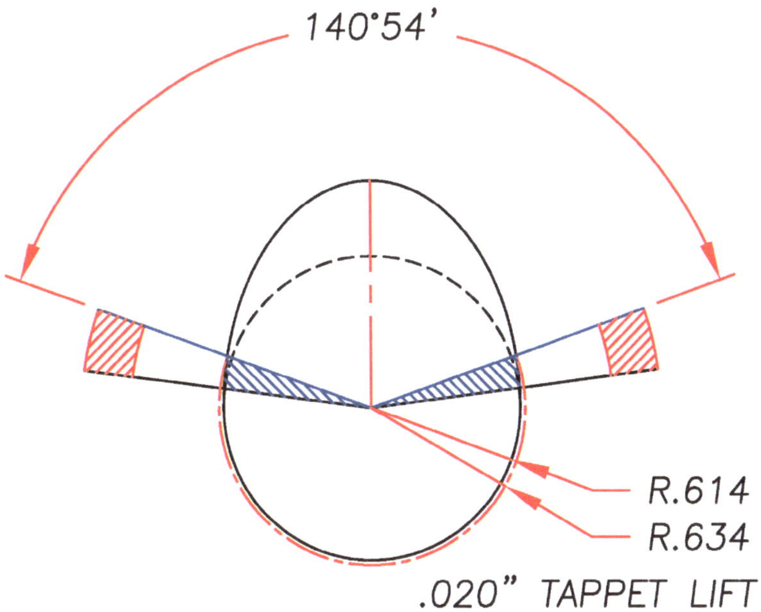

CAMS 101-C

VALVE LASH is shown here in the shaded areas, and our net running duration is 140°54'. Not to be confused, this is **CAM DURATION** (not crank) that is left over after measuring from a valve lash of .020". On a **CAM CARD** this would properly be read as DOUBLE this figure, or 282°.

The **CAM LOBE** is represented in **BLACK** with a base circle shown as R.614". The **RED** dashed R.624" represents the .020" float clearance of the tappet before the lash is taken up on the cam flanks. This .020" tappet lift duration used to be a common reference value, but so did about four or five other choices from .010" to .030" cam lift. Finally, most everyone settled on .050" as a relative standard to gauge real duration by; and most companies show a second higher value that is their "advertised duration."

Although I refer to the above .020" as "lash" (which many cam companies use), as I've said before, and will repeat again, I highly recommend **Valve Lash** be constrained to the tighter figures of .012" to .015" for the Intake valve, and no more than .015" to .018" for the exhaust. There is no design value to the arbitrary .020" (and greater) lash figures often used by cam companies. Tighter lash yields more stability, less impact loads and more active duration. The only caution is professional care for an accurate feel be made by the mechanic.

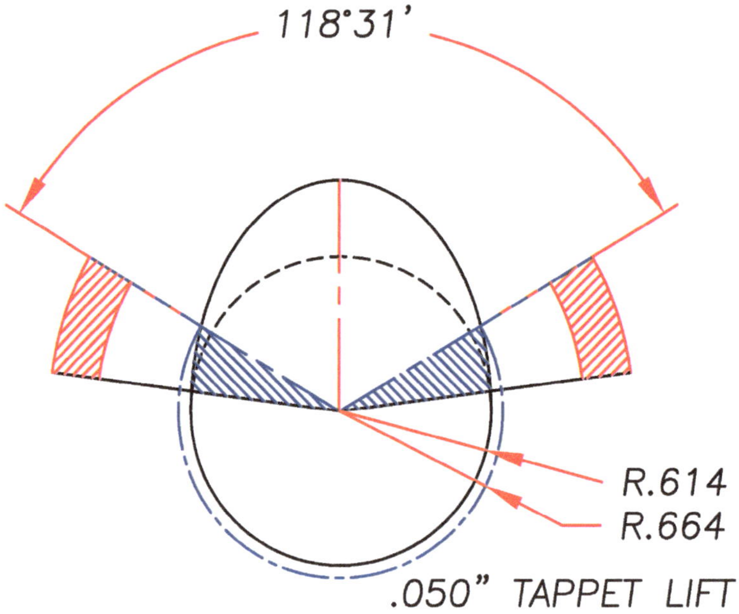

CAMS 101-D

DURATION @ .050" CAM LIFT is shown here as 118° 31' with the .050" portion of the cam flanks shaded to illustrate what a big chunk of cam duration they take. As noted, the duration on the CRANK (and cam card) would be shown as 237° @ .050" tappet lift.

The preceding illustrations of lesser values between .010" to .030" were randomly used by engineers until the .050" value was finally accepted by all. This is how cams are compared by all the popular OHV cam manufacturers, even though the second term of "gross duration" may also be given on the cam card. Gross duration is most relative on hydraulic cams.

The "standard" of .050" TAPPET LIFT came into being without any real cut in stone reason. One or two of the most popular cam companies of the 1960s and '70s era of cam development began using it, and the rest of the pack followed. One technical reason .050" is preferred over lesser values is because .050" was definitively ON THE ACCELERATION RAMP, and was NOT influenced by the slower initial acceleration rates often mistaken as the "lash ramp." It was important to get a real duration value that represented the cam's principal profile in this zone of acceleration. Why?

Because lower values of lift measurement are deceiving. Measuring a cam at .010" or .012" tappet lift exaggerates the duration by including the slower lash ramp area that is just coming off the base circle, and implies by this larger duration value that the cam is greater than it really is, compared to an .050" measurement. By choosing to measure the cam after the lash ramp's slower, initial acceleration rate has passed, we get a more accurate "apples-to-apples" comparison with other cams at this same .050" value.

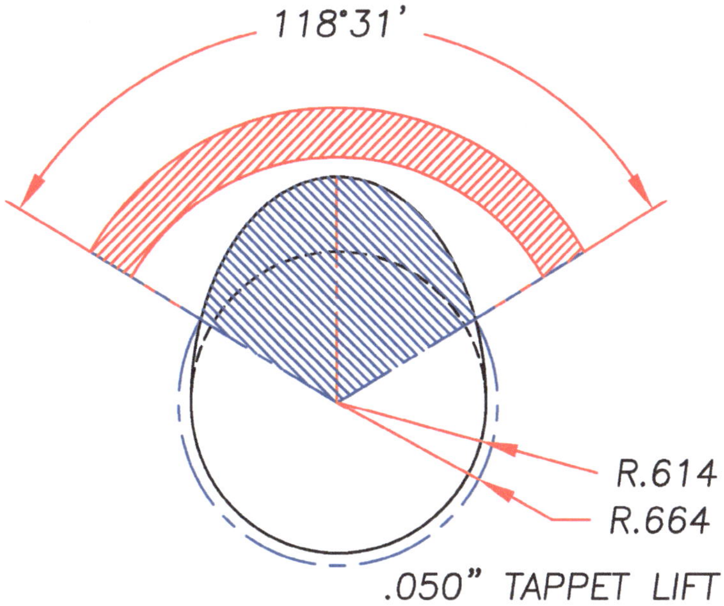

CAMS 101-E

SAME CAM @ .050" DURATION as shown in 101-D of the preceding page is shown here in the swept area only, with the hashed **BLUE** lines of the cam's base circle.

The R.664 (Radius .664") is on a CAD reference to indicate the tappet's intersection with the lobe, which has a radius of .614". The tappet isn't really "floating" .050" over the cam lobe, although some engines have found such extreme valve lash situations made power, for a short period of time. The acceleration ramp has already been entered by .050", and this just gives us a relative comparison to other designs by measuring it where the real strength of velocity is beginning to kick in; and well past the valve lash.

VALVE LASH

This is a good time to speak about valve lash, which I mention elsewhere, but is worth repeating. The valve lash settings will be "tight," for a cam that is properly designed, and properly matched for the package (heads, compression, cubic inches, and use). By tight, I mean .010" to no more than .016" for the exhaust. For many years valve lash was determined by the OEM and early racing era cam manufacturers to be around .020" to .025". There was reasons for this that are relevant to its real purpose, which is simply to allow for heat and expansion. That's it. Nothing more. The early OEM figures were pretty much a random, "sounds good to me" agreement, not founded in fact for what happens when this is unnecessarily too loose, as these figures on today's cams would be. Today's cams have more velocity, and the less "hammer effect" you can have by beginning the opening process sooner with less lash, as well as reap the benefits of another 4° or 5° of net duration, the smoother, quieter and longer living your valve-train will be.

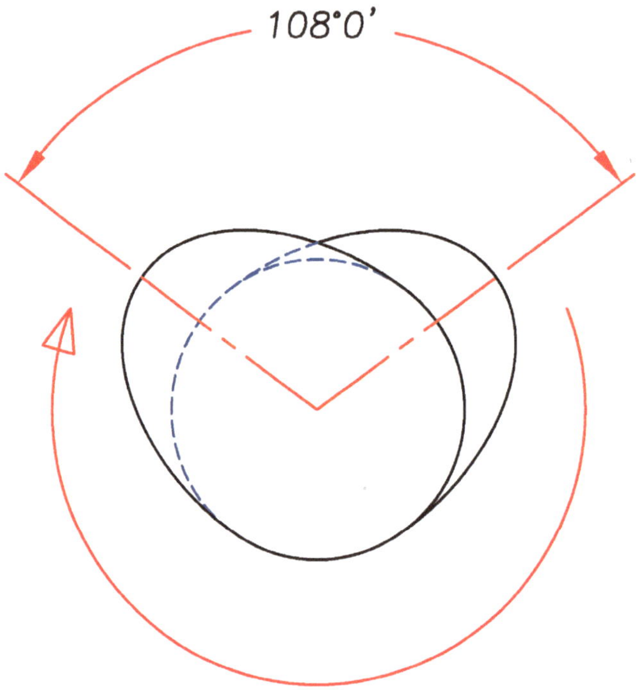

CAMS 102-A

LOBE SEPARATION is shown with the required second cam lobe now added to represent ONE cylinder of a V6 or V8 OHV engine; to create the indicated 108°.

The LEFT cam is the INTAKE while the **RIGHT** cam is the EXHAUST. The dashed **BLUE** circle illustrates the BASE CIRCLE as if continuing in between the lobes. In reality this is rarely the case, but here it is shown to indicated approximate beginning and end of the cam lobe flanks.

NOTE: CAM LOBE, **CAM BASE CIRCLE, CAM NOSE, ACCELERATION RAMP** (FLANKS) are illustrated in**:** **TERMS & DEFINITIONS** (p.111).

CAMS 102-A and **102-B** illustrates the relationship between different cams of a given "LOBE SEPARATION," which is a term often used in cam design and more often in sales and selection. But like many things with engine terms, it is actually a opposite perspective of what's really happening between the choices. The more accurate term that was traditionally used is "OVERLAP" (shown on the next illustrations).

What is worth paying attention to in the following illustrations is the effect on OVERLAP changes have when "Lobe Separation" is increased (widened) or decreased (reduced) in angular terms. OVERLAP is the more important "perspective" to think about, because that is the cause that changes the effect of what your engine's response at lower RPMs will be, as well as what compression ratio is needed to offset what is lost by increasing the overlap. Confused? Don't be. Increasing overlap is needed for high RPMs. But you lose "Net Effective Cylinder Pressure" (NECP), or the actual "running compression."

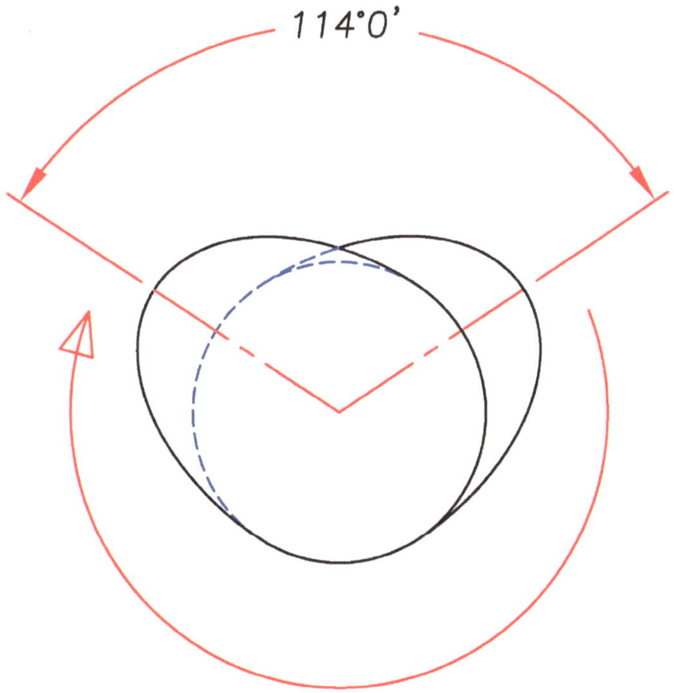

CAMS 102-B

LOBE SEPARATION here shows the cam spacing being INCREASED from 108° to 114°. As with preceding illustration.

This is a typical spacing value for lower compression engines seeking to boost their NECP if the OEM cam was 110°, for instance. More low-end torque would be realized, and usually so too would be better throttle response. This is *not* a lobe angle you would want with a high compression engine, or cheap gas. Because this higher lobe separation means lower overlap, the engine is starved sooner at higher RPMs, compared to a 110° or lower lobe separation cam.

The preceding illustrates a 108° LOBE SEPARATION, where here in **102-B** we have 114°. Compare the "V" that is created where these two lobes overlap each other, as you move from the preceding to the next illustrations.

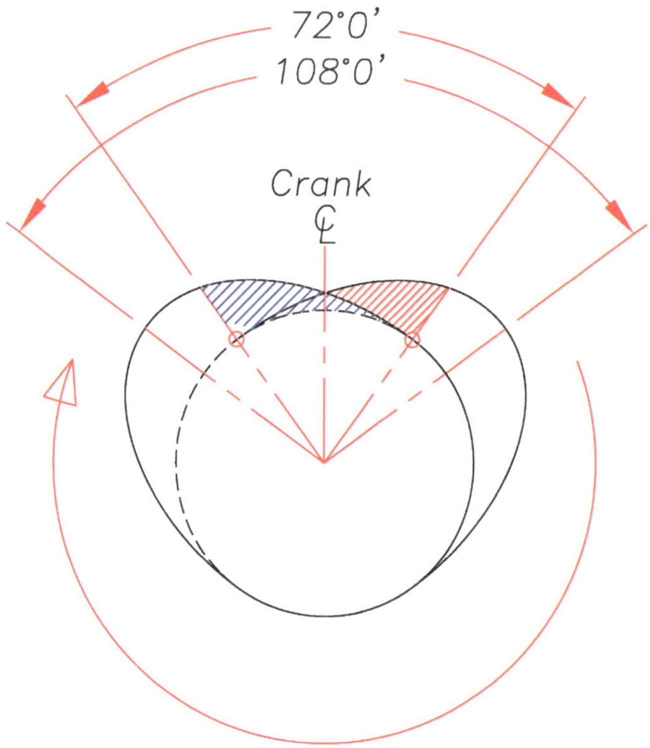

CAMS 103-A

OVERLAP and its effect is shown here with the SHADED area as 72° for our 108° LOBE SEPARATION.

The BLUE shaded area is the INTAKE cam, while the **RED** is the EXHAUST cam.

PHASING with the CRANK is represented by the vertical Center-Line (**CL**). This is the traditional symmetrical profile and its timing with the crank. Another way to think of this CL is TDC for the CRANK and PISTON.

Here in **103** you can see we've returned to the 108° Lobe Separation and depicted the OVERLAP at its 72° while shading in the active overlap zones where both intake and exhaust valves are open. See the effect on overlap of increasing the lobe separation in the next illustration.

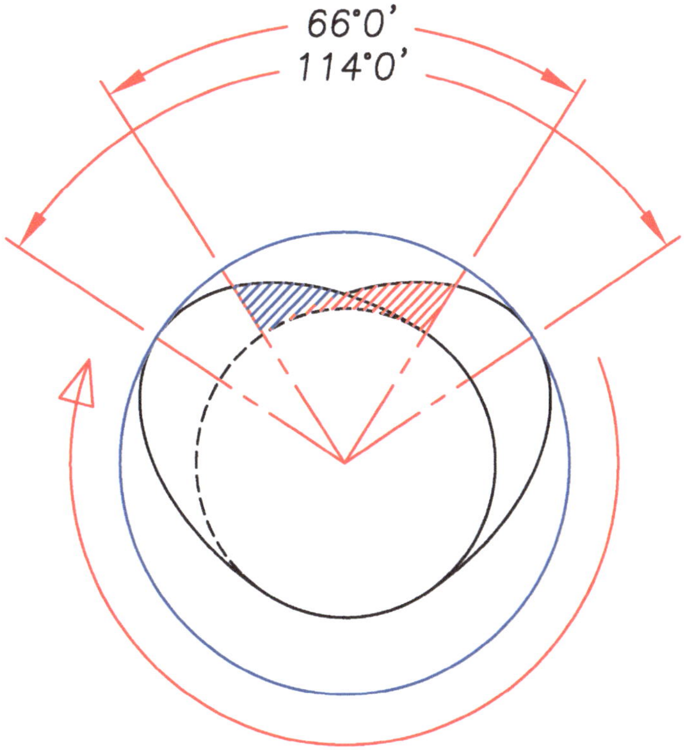

CAMS 103-B

OVERLAP and its effect is shown here with the SHADED area as 66° for our 114° LOBE SEPARATION. Illustrating how an INCREASE in LOBE SEPARATION creates a DECREASE in OVERLAP.

Here in **103-B** you can see that increasing from 108° to 114° Lobe Separation reduces the previous example's 72° OVERLAP to only 66°. This is what is really happening when you change lobe separation. Increasing Separation decreases Overlap, and the result is more **NECP,** or more simply stated, more "running compression." But the down side to this is that the engine loses upper RPM breathing capability. So in essence you trade upper RPM power for low end torque. This was a good remedy for the detuning of past high performance engines which suffered from reduced static compression ratio.

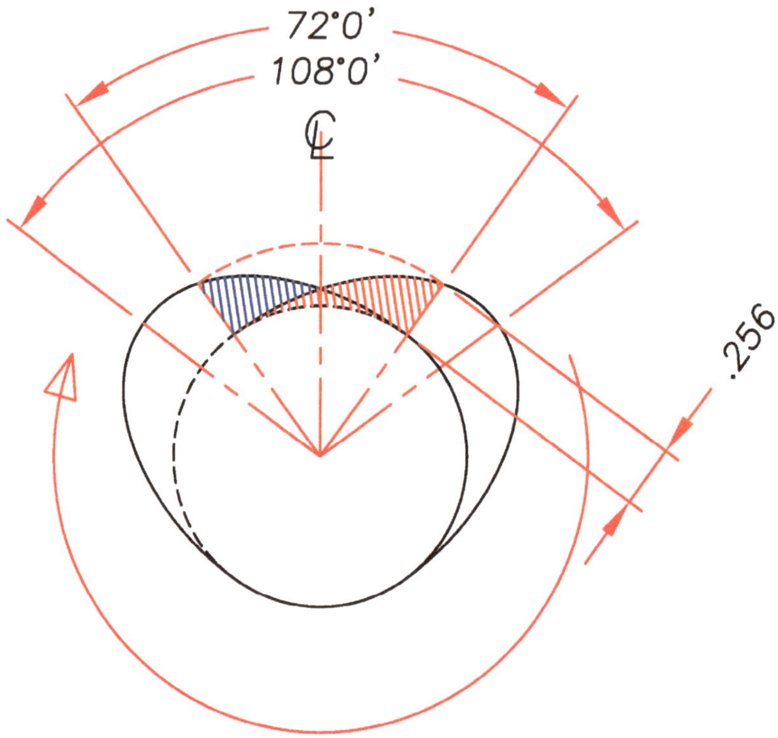

CAMS 104-A

OVERLAP and its effect is shown here with the SHADED area of 72° for our 108° LOBE SEPARATION being measured in CAM LIFT only to the point of .256". The **DASHED** line depicts the intersection points on the cam lobes where OVERLAP peaks. See the next illustration for the effect in cam lift at overlap with the increased LOBE SEPARATION.

Here in **104** through **105** we deal with the effects of Lobe Separation and Overlap upon cam timing with the CRANKSHAFT (Piston). You can see the physical influence that overlap has on Piston to Valve clearance by noting the .256" Overlap height depicted by the Purple Circle and value shown to the right.

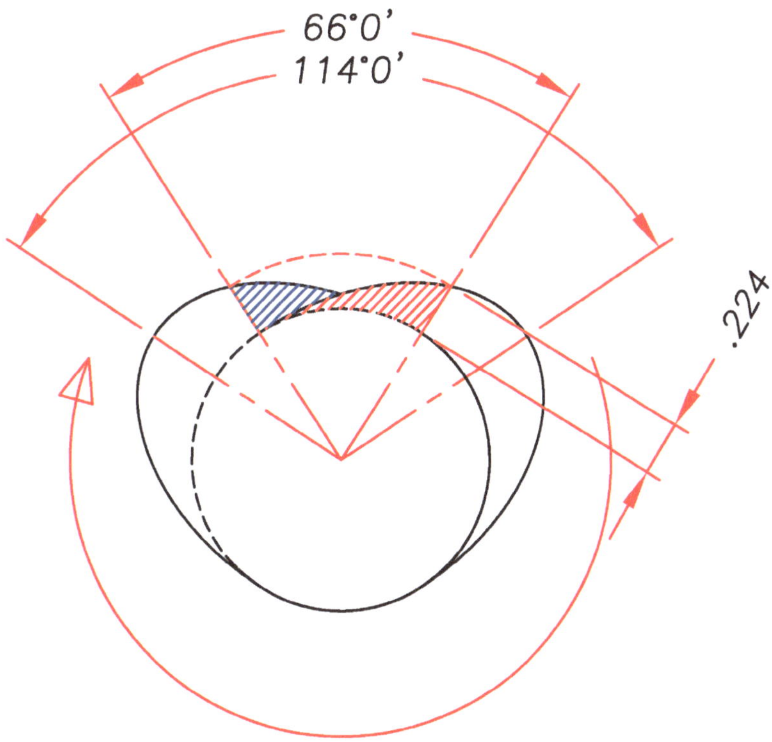

CAMS 104-B

OVERLAP and its effect is shown here with the SHADED area of 66° for our 114° LOBE SEPARATION being measured in CAM LIFT (only to this point) of .224". Considerably less. **PHASING** with the CRANK in this and the previous illustrations is still symmetrical, although absent the Center-Line shown in **103**.

Here in **104-B** the .256" Overlap Height is reduced to .224" in addition to being only 66°. This would relate to being equally divided with the crankshaft's TDC, known as "straight up." But to gain optimum performance from this equally divided illustration, where you may need to advance or retard the cam with the crank.

Advancing or retarding the cam shifts where the valve moves closer to or farther from the piston. Of the four valve timing events as measured to the crankshaft position, the most important is the INTAKE'S CLOSING point. When you advance or retard a cam, what you're really doing that influences the engine the most (specifically the heads) is moving where the intake valve closes in relation to the piston. That is the dynamic the torque curve, peak RPM and throttle response are all responding from. The other three valve timing events (Intake opening; Exhaust opening; and Exhaust closing) are almost insignificant by comparison. The Intake Closing point is on the cam's opposite side of the overlap cycle. So when you choose a centerline angle that is less (greater overlap), you're increasing the scavenging time of the exhaust cycle. To keep the engine throttle response low with this greater overlap you would ADVANCE the CAM to compensate; which is restoring the intake's closing point with the piston that was moved by increasing your cam centerline angles. The static (paper) compression needs to be higher for cams with greater overlap because the NECP bleeds off compared to a cam with lower overlap.

(See: LOBE SEPARATION, pp.70, 83 and 121.)

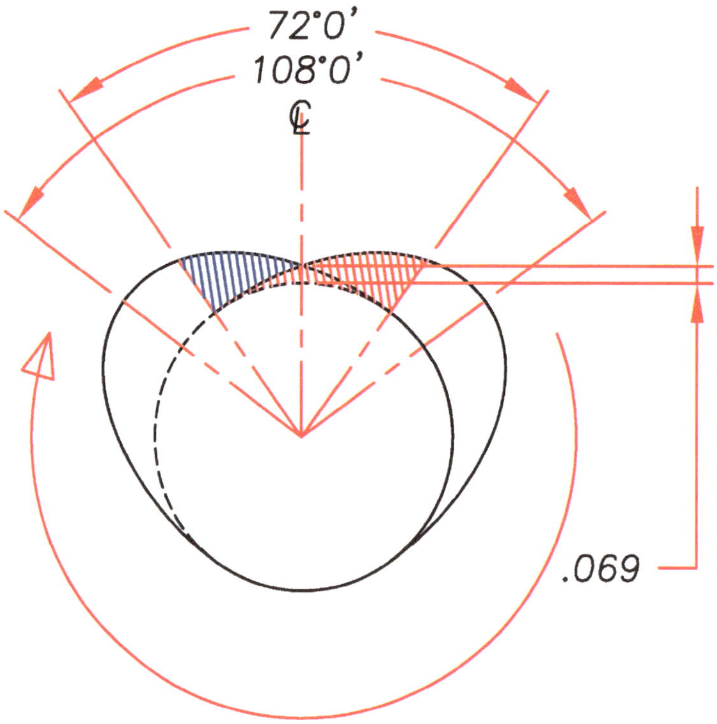

CAMS 105-A

OVERLAP and **PHASING** "straight up" as depicted by the CL, base CAM LIFT is shown as a mere .069". The **PURPLE** circle depicts the intersection points on the cam lobes where OVERLAP meets. See next illustration for effect in cam lift at overlap with the increased LOBE SEPARATION.

Here in **105-A** we use another perspective to show an important point which directly affects the piston to valve clearance that occurs when both valve are open. Here the Purple Circle shows the HEIGHT of the INTERSECTION as .069". But as mentioned in the previous illustration, this is a "STRAIGHT UP" **cam phasing** sequence is ground zero for most cam development starting points with various engine combinations that ultimately get "advanced" or "retarded" to fit the RPM band that is desired for that engine's particular application. But that is too simple. Cubic inches, compression ratio, fuel, cylinder head volumetric efficiency and driveline are all essential investors in a final cam's best design.

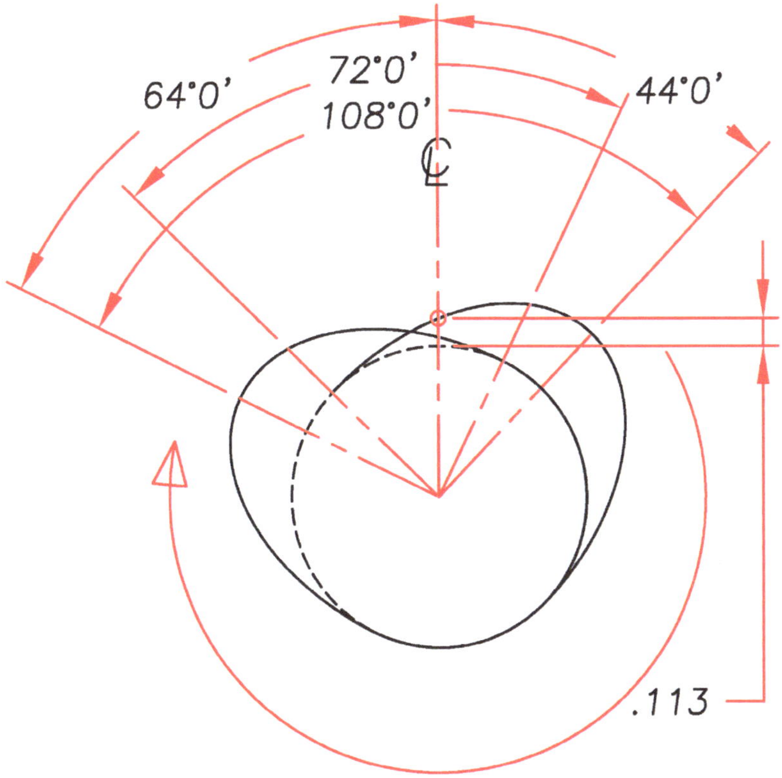

CAMS 105-B

OVERLAP and **PHASING** "retarded" as depicted by the cam rotation, shows an increase to .113", or .044" more than the "straight up" phasing. An exaggerated shift to retard the cam 10° has been used to show this effect. Even though you would never time a cam this far behind in crank degrees (20°), the cam manufacturer might. This .044" increase would translate to a valve lift increase of .066" with a 1.50:1 ratio rocker arm, and clearly **PISTON TO VALVE** (**PTV**) would definitely need to be checked.

Here in **105-B** we've retarded the cam an exaggerated 10° – which in "crank" terms means 20° – a value beyond what you would ever really use, but for illustrating the point is better made. Our previous Overlap Height was shown as .069" and now it is .113". You may not think this is significant, until you remember this is CAM LIFT specs, which will be multiplied by ROCKER RATIO. So a 1.50:1 Rocker arm will multiply this increased value of .044" to .066" at the piston. Even taking this exaggerated view into account by half as much, that additional .033" increase could easily be the difference between having enough piston to valve clearance or NOT. I hope this sequence helps some of you get a little more from what you choose.

At the engine builder level it is nearly impossible to predict what the NET piston to valve clearance will be for a given profile that has **not** been tried before in a specific engine combination (unless you have solid modeling software). When you move the cam around, or even go to a *smaller* (less lift) cam, it can still DECREASE your PTV, because of lift velocities on the ramps. Measuring or comparing two cams of the same .050" duration, where you know what the PTV is with one, does **NOT** guarantee the same clearance on the second design.

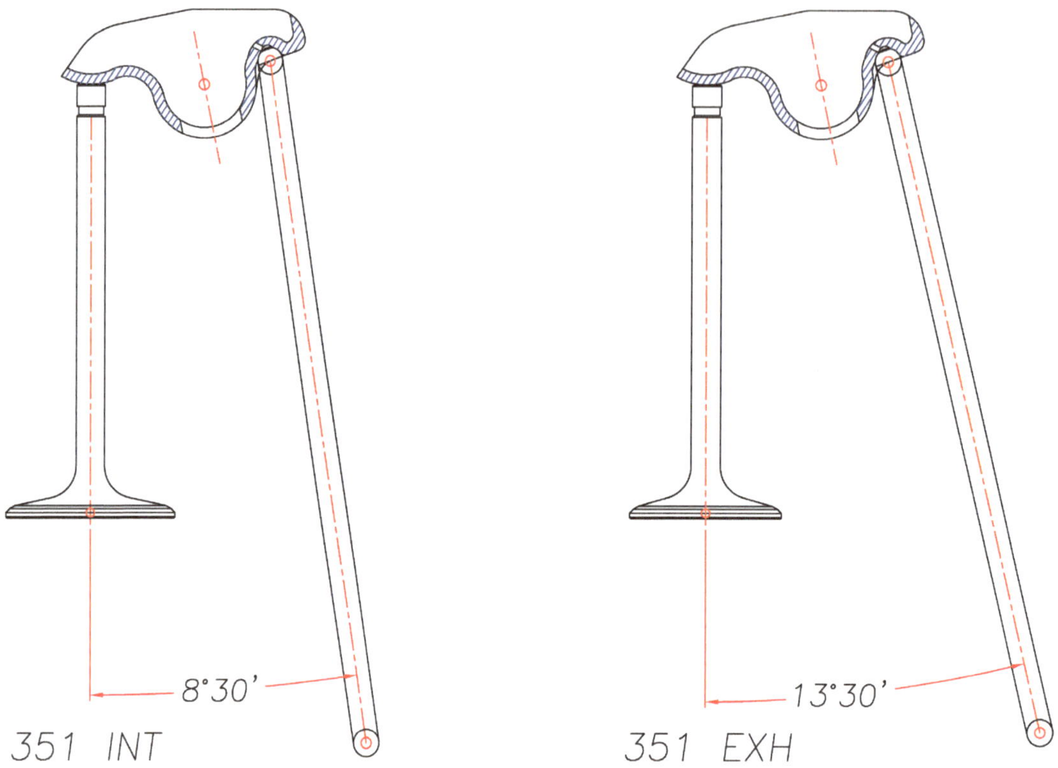

FORD'S PRECEDENT

351 CLEVELAND

The 351 Cleveland requires separate rocker designs for the intake and exhaust, to be really precise. The ***Design Geometry*** above is 8.5° Intake and 13.5° Exhaust. On the next page you'll see the 429/460, and my reference to the BB Chevy, which in rocker land is often interchanged, poorly.

What's so "precedent" about Ford's design on this and the next page is they use **the same rocker arm**. Think about this for a moment: Above you see an 8.5° Intake attack angle. On the facing page you will see the Big Block Chevrolet has a 23.5° Exhaust. These are the two extremes, and as such they have a 15.0° spread between them. So what? *The rocker arms only move 24° to open the valves!* (+/-) If the same rocker arm is used, and is 15° WRONG between one extreme and the other, how can you expect the aftermarket industry to drill down on precision geometry? Remember, both of these engines come out of the 1960s, way before the level of cam and valve-train technology we had only ten or twenty years later. But this is the lengths that a big company would go to, to save a buck!

Numerous companies have made rockers for the 351 Cleveland, but most of these are actually their BB Chevy design with no difference other than a misleading number on the box. The *problem* is, these brand name companies won't reveal the geometry specs to engine builders; and except for a couple companies who've copied Mid-Lift, they retained their old design geometry of 25° to 30° or worse. So when used on the Cleveland 8.0° and 13.5° geometry, they're already 20° too high. Parts breakers.

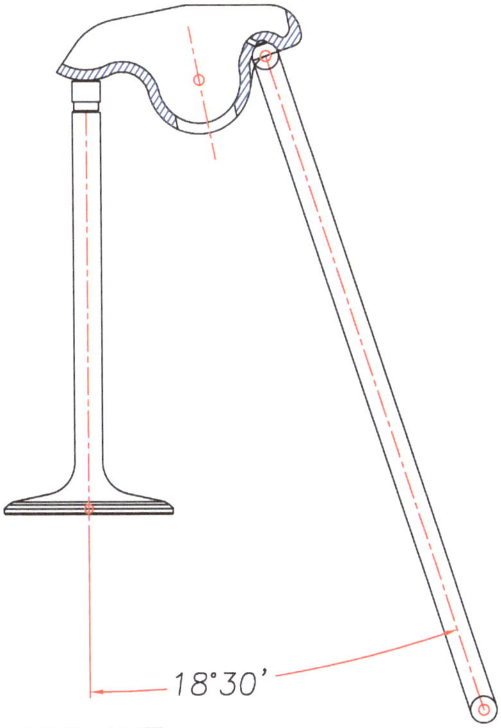

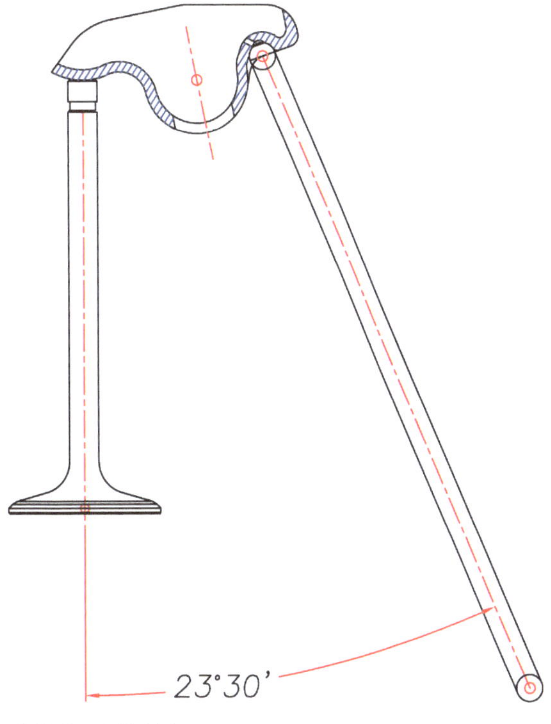

FORD'S PRECEDENT

429/460 385 SERIES

The 429/460 Ford also requires separate rocker designs for the intake and exhaust, to be really precise. The ***Design Geometry*** above is 18.5° Intake and 23.5° Exhaust. The 385 Series engine is much closer to the Big Block 396/454 Chevrolet than the 351 Cleveland shown on the previous illustration. But here too you can see there is a 5.0° difference between the intake and exhaust Pushrod Geometry.

The BB Chevy has an 18.0° intake, and a 20.5° exhaust pushrod geometry. Obviously, only a 2.5° difference. Unlike Ford, who took the cheap way out on their blocks with in-line tappet bores, Chevrolet laid their exhaust tappet angle over 4.0° from the intakes, to compensate for the exhaust valve's inch or so offset from the intake valve, away from the intake manifold. This improved the pushrod geometry with the tappet for where the exhaust rocker needed to be. On the PVS BB Chevy "45 Series" PA4570 (and 1.8:1 ratio PA4580), I used the SB Chevy angle of 19.0° because it averaged nicely between the two BB Chevy valve angles. But because the stud to valve distance is close between the Ford 351, Ford 460 and BB Chevy, the BB Chevy rockers are often used on all three engines, as noted to detriment for the 351C on the facing page. Against my recommendations, hundreds of sets of PVS PA4570 rockers were bought for use on the 351C, especially *down under* in Australia and New Zealand, with high acclaim, despite the rocker's pushrod geometry being 19.0° for an engine needing 8.0° and 13.5°. The 5.5° and 10.5° deviation was acceptable by knowledgeable engine builders who understood the tradeoffs.

Ford's Precedent more rightly belongs in ***The History of MID-LIFT***, to justify the rocker confusion.

CAM DEFINITIONS

INTRODUCTION

NOTE: *Some of the terms in this chapter can also be found in TERMS & DEFINITIONS (p.111), mixed with other terms by their alphabetical order. This chapter consolidates them to simplify a dedicated study of camshaft lingo, which also includes additional TIPS relevant to the CAM BASICS Chapter.*

CAM: An eccentric shaped circle that has a base diameter which is less than 360° that rotates about an axis which increases its diameter from the axis as it rotates through a 360 degree rotation. Generally egg shaped on designs that work with a direct contact running surface "cam follower," or having more squared radius` for designs where the cam follower incorporates a roller ended running surface. **See: CAM FOLLOWER** (pp.85 and 120).

CHOOSING A CAM: The following are the basic fundamentals relevant to selecting a cam. It is by no means a cut in stone blueprint where some of these variables can't be stretched. But this should keep your feet on the ground to avoid losing perspective of what is important, over what someone is hyping to make a sale, or worse, failing to give you what you really need.

> There are THREE components of cam design which must each be chosen for your engine. These are: **Valve Lift, Duration, and Lobe Separation**. When speaking directly to the cam company of your choice for a custom profile, either you or they will translate a fourth important value of the equation into "cam talk" and that is the Rocker Ratio. Because the cam company (and you) should be thinking in terms of "Lobe Lift" on the cam itself, to avoid a changed rocker ratio from being missed in the discussion; easily done. **See: CHOOSING A CAM** (p.99).

The Chapter **CHOOSING VALVE LIFT** (p.97), will give you the fundamentals to avoid the first mistake engine builders often make in choosing a cam. You must know the REAL value of your cylinder heads, because this is what your valve lift is really attached to.

DURATION: Is a value that is spoken of in a couple different ways, usually Gross and "050," (O-Fifty). The .050" tappet lift is what you want to focus on, and not pay much attention to the Gross, or any other value of tappet lift (i.e., .020"). O-fifty is the standard everyone has finally settled on for the last 30 plus years. Here's the fundamental point with DURATION. *The higher the RPMs, the longer the valves need to be held at a given lift to accommodate giving the head time to respond with additional airflow*. At lower engine speeds, having the valves stay open longer than they need (for that RPM) will show up with sluggish response. So we are making choices that are always a compromise aimed to enhance the major area of use.

> Duration is relevant to TWO values: CUBIC INCHES of your engine, and RPM to be used. To confuse you even more, both of them offset each other; and not taking that into consideration can get you into trouble, or at the very least – leave you short of what you require in this department. Without making a big deal out of this, let me just repeat a small point for comparison: If you have a large cubic inch engine for the "family" size it comes from (i.e., a 430 cubic inch Small Block Chevy, who's "family" is 302–350 CI), then this engine has larger demands for cylinder head requirements (port size) than it was originally designed. But most important, is this increased size obviously came from an INCREASED STROKE.

CAM DEFINITIONS

Increased stroke engines will have the piston operating at a faster speed, "per RPM," than what this engine originally had its components selected for. The engine "thinks" it is running 7,000 RPM when the tach only says 6,500. Accordingly most "stroker" engines don't turn as many RPMs as their smaller counterparts. However, they need the same duration cam. So with stroker engines, don't let the fact that you know you'll be turning less RPM detour you from a duration that a smaller engine turning more RPM would require. A Big Block engine was designed from a family of larger cubic inches to begin with, so the term stroker for this family usually denotes 500 cubic inches and up, much up in some cases, topping 700 cubic inches.

LOBE SEPARATION: Is – like many engine terms – used to describe one aspect that is actually doing something else. As mentioned below under this term, this determines **CAM OVERLAP** (the period when both intake and exhaust valves are open simultaneously). This overlap is what determines *Net Effective Cylinder Pressure* (**NECP**), or the actual running compression your engine sees, as measured in pressure. This is different than the theoretical pressure attained from your **STATIC COMPRESSION RATIO**, which is derived from dividing the net volume of your combustion chamber by the total volume of your cylinder, as measured from perfect Top Dead Center (**TDC**) and Bottom Dead Center (**BDC**) of the piston. That has no real value for determining NECP because the cam's overlap is the final dictator of that. *The more cam overlap you have, the less running compression you have.* High RPM engines need more OVERLAP, so the valves can be open together longer, and thus "purge" the wasted gases. But the price for this longer overlap is that we LOSE effective (running) compression. So to recover that loss, we increase the STATIC compression ratio. *This order is an important point. Changing compression ratio without understanding the order of priorities it requires can allow the other components to be overlooked, making for a hot engine.*

Here's the fundamental RULE with **LOBE SEPARATION**. *The higher your STATIC COMPRESSION RATIO, the LOWER your LOBE SEPARATION.* The range of lobe separation most often seen is between 104° and 114°. As noted in the definition for lobe separation (above), this is the actual physical angle between the cam lobe centerlines of the camshaft. Not to be confused with crank degrees, as most all other cam information is referenced to.

So here's a yardstick of REAL numbers if you want to skip past all this background information: If you have a naturally aspirated engine (which is what all these recommendations is aimed at), and you have a WEDGE style cylinder head (i.e., SB or BB Chevy or Ford) with 8.5:1 to 9.5:1 CR (compression ratio), your cam should be around 112°–114°. If you're CR is up around 14:1, it should be around 107° to 108°. If you're more than that, then your cam's Lobe Separation could be best at 106°.

If you have a HEMI COMBUSTION CHAMBER, then all of the above figures will drop by 2° to 4°. The highest Lobe Separation value you'd want on a Hemi (Chrysler or Ford BOSS) is 110° and the lowest you'd have is 102°, depending on your compression ratio. A 13.5:1 CR likes 104° Lobe Separation on most American designed OHV Hemi engines; while lower compression ratios would prefer 108° and 110°.

Any compression ratios you have that fall in between these outside examples should be adjusted evenly between their spread. For instance, if you have 11.5:1 to 12:0:1 CR, then 108° to 110° would be a safe choice.

CAM DEFINITIONS

NOTE: *An additional dynamic that should be considered in the above, as well as with duration choices, is that cam ramps have varying velocities from one company's design to another, and for one proposed mission versus another. This difference is NOT quantified in the cam company's CAM CARD. A cam with a quicker ramp (more velocity at a given point on the opening ramp), is in essence a "bigger" cam than a like duration version which is more moderate on the acceleration rate. In such cases, you would treat the lobe separation and/or the duration chosen from this insight. In other words, if a cam is very aggressive, then the lobe separation might need to be tweaked to a 1° or 2° higher number because the overlap is greater at the middle of the cam ramp area, where you're not measuring from. Your cam manufacturer should be able to fine tune their recommendations for you when you are getting this nitty-gritty, but you won't get such answers from the usual "call center" kid answering phones and reading from an advertisement flier. You need to speak to the cam engineer, or their assistants who are up on these differences. To CHECK that you're getting the right info, or just to know what you're comparing, you need something like a Cam Doctor. Once you understand how important this is, from having done it, you'll never look at your engine building choices the same way again. If you have, or often use a Dyno, this should be SOP (standard operating procedure).*

> **REMEMBER:** When you're talking about cam designs and comparisons, you're doing it from a fixed point on the cam lobe, which is the point where the tappet has been lifted from the base circle by .050" of an inch on the opening side, until it returns to .050" of an inch on the closing side. *You can have TWO cams of identical durations as measured from these two points, while even having the same CAM LIFT, but have entirely different event velocity in between these points, and thus, an entirely different affect on the engine.*

CAM BASE CIRCLE: (a.k.a. The "**HEEL**") The lowest point of measurement to the axis of rotation, this is the constant radius, NON-eccentric running surface of the camshaft that has NO effect on lift. It is also the mathematical foundation, as measured in degrees of rotation, for the starting and stopping points of where the lash ramps and acceleration ramps begin and end. However, these specific measuring points are gauged from a relative point, marked as a value of lift from the instant tangent point of where the ramps meet the true base circle. These points have been different over various cam companies and years of cam evolution, where tappet lift measurements such as .010", .012", .015" , 020" and so on were used to describe "gross duration." But the "standard" by which most real comparisons between cams is made, is marked at .050" tappet rise off "base circle." This higher tappet lift measuring point more equally compares two cams without the influence of lash ramps, which often can stretch out to look like a big cam, when measuring in this gross duration way, but in fact may be very tame by comparison to their .050" figures.

CAM CARD: The printed record of cam specifications provided with a camshaft to give fundamental design information necessary for selection, use and installation within an engine. This information should include **CAM LIFT**, **CAM DURATION** (often shown in two values, but most important is the .050" Tappet Lift), **CAM LOBE SEPARATION**, **CAM PHASING** (installed timing with crankshaft) and **VALVE LASH** (if mechanical). Additional information often included is "Theoretical" VALVE LIFT (based on specific rocker ratio, often not accurate to reality) and "Theoretical" DURATION (often taken at irrelevant lifter values close to the closed valve point, such as .012", .015" or .020" Tappet Lift; and usually of most value to hydraulic grinds, but often not considered of much value to experienced engine builders).

CAM DEFINITIONS

CAM CENTERLINE: Actually has two meanings. The usual meaning refers to the cam's designed INTAKE centerline position, in CRANK degrees, as measured with the CRANKSHAFT position AFTER TOP DEAD CENTER (**ATDC**) of the piston. It is also known as the specification for which the cam's "**PHASING**" is used. Again, this is the advancing or retarding of the camshaft with the crank's timing (piston position to TDC or BDC), per the cam manufacturer's original design specifications. (See: **CAM LOBE SEPARATION**) The second meaning, pertains to the more obvious implications of its name, the actual centerline running through the base circle's axis, up through the cam's nose, thus dividing the opening and closing events equally. This can be misleading since many **ASYMMETRICAL** cams don't have nose specifications that are evenly divided. None the less, there is a theoretical centerline to the computer's data which calculated and assembled the **CAM RAMP** dynamics.

CAM, DUAL PATTERN: Refers to different Intake and Exhaust Cam profiles. In the old days which stretched up into the early 1970s, camshafts commonly had identical Intake and Exhaust Lobe profiles ("single pattern"). Although custom grinds of dual patterns had been made long before the 1970s, it wasn't common. For the last thirty years, it is more uncommon to find a performance cam that isn't dual pattern than it is to find one that is.

CAM DURATION: This the period of "time" in which the **CAM LOBE** is active beneath the **CAM FOLLOWER**, as measured in one of two ways: degrees of rotation at the cam, or degrees of rotation at the crankshaft, the latter being the more preferred standard with regard to engine builder terminology. Cam Duration historically has been measured from various, non-conforming reference points of tappet lift for its "*gross duration*," which technically is the total duration measured from the instant tangent point of cam ramp departure from, and returning back to the **BASE CIRCLE**. Gross duration has also been referred to on cam cards in different ways by various cam manufacturers who choose various starting (and ending) points for where this measurement is taken, always termed in thousandths of an inch rise from base circle; such as .012", .015", .020" or more of tappet lift on both opening and closing sides of the cam. The more this figure becomes, the less "gross" it is. So these higher measurements were more often used for timing purposes to install the cam with the crank by, or to compare one design to another for the effective duration (actually seen by the cylinder heads). But thankfully this comparison reference point has settled down in the last 20 years or so to including the more preferred .050" TAPPET LIFT reference. From .050" (fifty-thousandths of an inch) LIFT from BASE CIRCLE on BOTH Opening and Closing sides of the cam lobe, CRANK degree measurements are used to mark a particular cam's most active profile. To get some perspective on selecting cam duration for your engine, **see: CHOOSING A CAM** (p.99).

CAM FOLLOWER: (TAPPET or LIFTER) A cylinder shaped component that rides upon the cam lobe of an **OHV** engine, and accepts a **PUSHROD** on the opposite end which provides the reaching link to the rocker arm atop the cylinder head. Cam followers are either hydraulic "roller," hydraulic "mechanical," or mechanical roller or straight mechanical, a.k.a. "SOLIDS." The ultimate performing cam follower is a SOLID ROLLER (no hydraulic valve lash preload); often simply referred to as a "roller." A roller lifter uses a hardened roller race which runs directly upon the cam lobe in conjunction with hardened needle rollers to create a true needle bearing of high strength and impact resistance. (See: p.85)

This bearing wheel is permanently mounted within the tappet's housing with an interference fit solid pin, that may or may not have additional locking rings retained in grooves for added security. The ROLLER Tappet requires a dedicated CAM lobe, for both the hardened surfaces required for mating to the hardened roller's race.

CAM DEFINITIONS

But most specifically in corresponding with the angular acceleration formulas which a cam ramp mating to a constantly rotating tangent point requires to keep the acceleration dynamics of the tappet "in-line" with the acceleration rates pushing up from the cam centerline. The simple way to understand this, is that the pressure point of a cam lobe contacting the roller is shifting around the roller as the roller follows the cam lobe up and around the top (cam nose). Whereby a SOLID tappet which has a "fixed" surface laying directly atop the cam lobe is always receiving its upward lift in a straight line parallel with the tappet body, away from the cam's centerline.

> **NOTE:** The velocity of the flat tappet lifter/cam is limited to a specific value by the DIAMETER of the tappet body and the BASE CIRCLE of the cam in conjunction with the LOBE LIFT being used. Since the rubbing surface on the TAPPET'S bottom side receives its "lift" by the nose of the cam sweeping across most of its entire face (contact surface) in the process of rotating beneath it, if the LOBE LIFT is too great or has too much angle coming off the BASE CIRCLE in its attempt to create more velocity, then the LOBE literally sweeps outside the diameter of the tappet. Accordingly, the tappet is no longer running atop the cam if this has happened, and destruction occurs.
>
> The ROLLER tappet does not have this problem because the point of contact between the cam lobe and the roller is always moving around the axis of the roller tappets roller centerline as it follows up and down with each rotation of the cam lobe. This concept allows nearly no limits beyond those established by the laws of inertia, mass (weight) acceleration and harmonics reaching their limits first.

MUSHROOM TAPPETS (slang) are a variation of the traditional, one diameter body dimension that rides within the block's tappet bore, with the exception that the bottom "face" diameter was increased to a larger, "stepped" diameter that gave rise to the term "mushroom." This is a benefit only where the increased cam velocity and/or duration is needed for a particular engine's cylinder head, rpm and cubic inch combination; and where the ROLLER tappets are prohibited by class rules. Mushroom tappets were developed so their larger contact diameter would allow more aggressive cam rates (faster) for the base circles being used, which are always limited to the engine's particular cam bearing diameter. In point of fact, the mushroom lifter goes back many decades, where it was used on older and simpler engine designs (not requiring high cam rates), that simply wanted the lighter weight, and/or a smaller diameter tappet body, while still incorporating a larger contact face diameter to function as needed on the cam lobe. Racing applied the concept to the conventional design many years later as a loop hole technique to get around "non-roller" tappet cam rules for racing which required that only a flat tappet cam be used, particularly STOCK CAR racing (as it used to be called). Today, some classes of racing allow the mushroom tappet, and some don't. On the more popular Chevrolet and Ford engines where such detours in class rules allow, but prohibit the Mushroom tappet, the Chevrolet tappet bore of .843" is often bored out for the larger .875" diameter Ford tappet to accomplish the higher cam rate goals. This mere 1/32" can make a world of difference to increasing the cam rate.

CAM HEEL: See: CAM BASE CIRCLE (p.119).

CAM LIFT: This the actual measurement of the cam lobe's total rise from the **BASE CIRCLE** of the cam which is imposed on the **CAM FOLLOWER**. Cam lift on OHV engines is always a lower value of net **VALVE LIFT** because **ROCKER RATIO** multiplies this by 1.5:1 or more in most cases.

CAM LOBE: The eccentric portion extending from the cam's BASE CIRCLE to create CAM LIFT which comprises all three (3) stages of cam definition: LIFT, VELOCITY and DURATION, and also includes the transition ramp referred to as the LASH RAMP, which technically does not exist in of itself.

CAM DEFINITIONS

CAM LOBE SEPARATION: Pertains to the physical separation of cam lobe centerlines, in CAM degrees. This is also the reverse perspective of "OVERLAP." Both terms deal with the physical timing between the intake and exhaust lobes of a camshaft for the given cylinder in question. Traditionally, **OVERLAP** was the term used to deal with cam tuning to cylinder heads and cubic inch needs on a given cylinder head. In certain circles this still is. But LOBE SEPARATION became prevalent in the late 1970s as lower compression ratios became the norm from the Detroit manufacturers, adapting to more stringent EPA regulations. Increasing LOBE SEPARATION (which decreases overlap) is a way of building higher cylinder pressure on lower compression engines. See: Net Effective Cylinder Pressure (**NECP**), p.135. For perspective on cam lobe separation, **see: CHOOSING A CAM** (p.99).

CAM NOSE: The highest portion of the cam lobe eccentric where valve lift is at its greatest.

CAM OVERLAP: Refers to the time when both, the intake and exhaust valves of a given cylinder are open at the same time, measured in CAM DEGREES. Specifically, it is when the INTAKE valve begins to open, as the EXHAUST valve is just closing. The piston is still rising UP, chasing the exhaust valve closed, and the intake valve is opening to meet it. Increased overlap is needed on higher RPMs to provide additional time for the cylinder to "purge" itself of leftover gases from the prior combustion process. Physically, as explained in **LOBE SEPARATION**, the overlap cycle is the opposite perspective of the same consequence to cam design. They effect the engine's "running" **COMPRESSION RATIO**. You are merely saying the same thing from opposite perspectives. OVERLAP is a term not used much anymore, but was very much the perspective of cam development and comparison by professional engine builders during the pivotal cam development era of the late 1960s and 70s, where many of the breakthroughs of cam design were established. The transition to "lobe separation" lingo was the result of marketing by cam companies, beginning in the mid to late 1970s, who needed to boost cam sales and horsepower on LOW compression engines, resultant from the change in Detroit's gas guzzling car production, and ending the Muscle Car era as we knew it. **REDUCED** CAM OVERLAP is **INCREASED** LOBE SEPARATION. This reduced cam overlap increases **NECP** (**N**et **E**ffective **C**ylinder **P**ressure) within the cylinder, often boosting factory engine horsepower by in effect increasing the running compression on these milder, low compression engines. The greater OVERLAP was not needed on these low compression engines since they were not going to operate at high RPMs. Among advanced cam analysis in professional racing engines, cam overlap is still a critical perspective and term to be documented and used, particularly with cylinder head flow analysis.

> **NOTE:** It is important to understand that you need to increase the static compression ratio of pistons (i.e., 9:1 up to 12:1) because a cam of greater overlap will lose "running compression" (NECP). Increasing the piston's compression ratio replaces this lost operating pressure from the increased cam overlap. (Assuming you have cylinder heads capable of breathing at these higher valve lifts and RPMs.) You cannot increase cam overlap and duration on a low compression engine without changing the pistons for higher static compression, without losing performance. Nor can you install larger compression pistons without changing to a cam of higher overlap (and duration), since you will now have too much "running compression" (NECP) and likely burn pistons, valves and create other problems from excessive heat, pre-ignition and detonation. These two rules run hand in hand. The only exception to these necessities being chosen only for higher RPM operation would be a large cubic inch engine with a large stroke crankshaft; as this increased crank stroke increases **PISTON VELOCITY** and increases airflow through the heads at lower RPMs. Too often this is mistaken as torque from crank leverage.

CAM DEFINITIONS

CAM SPLIT-OVERLAP: (Camshafts) Refers to a point on the cam between the intake and exhaust lobes where the tappets (and valves) are lifted the same amount simultaneously. This is a confirmation reference point for timing with the Crank, once the integrity of each cam profile has been confirmed.

CAM RAMPS: Opening and Closing, as they are known, are the connecting portions of the cam's running surface that lies between the **BASE CIRCLE** and the **NOSE**. Sometimes referred to as the cam's FLANKS. The cam's **LASH RAMP** is also included here and designed to make the transition between the base circle and the more aggressive (quicker opening velocity) **ACCELERATION RAMP**. It's important not to get too distracted by these, because to the engine builder they really don't matter. The only thing that is important to cam selection is *duration; lobe separation* and *lift.* In that order!

CAMSHAFT: The multi-lobe shaft consisting of the required number of eccentric cam lobes needed to operate the valve-train, as it is timed by a chain or belt with the crankshaft. It is made of a variety of materials, with cams of 8620 alloy steel being the most common for roller tappet designs. Necessary of course for the hardened roller. High tech coatings and other manufacturing methods are changing this norm on certain applications. NASCAR breaks the rules on cost and technology in this regard.

> **CAMSHAFTS** of today's high performance OHV (Over-Head-Valve) engine technology are the "heart" of a FIVE (5) component system that makes up the **VALVE-TRAIN**, which is arguably an antiquated principle of design, dating back nearly a century. Aside from the last component in our list: the "VALVE," these other three components are respectively known as the LIFTER (or tappet, or cam follower), the PUSH-ROD and the ***ROCKER ARM***.

CAM VELOCITY: Is a term often used by cam companies to describe (in a general way) the cam lobe's rate of lift over other profiles. It of course has a very specific engineering value to cam designers, which is given a value of thousandths of an inch *per degree* of either cam rotation, or crank rotation (double the cam value). "RATE of ACCELERATION" is another general term used for the same meaning, usually pertaining to cams; but we often use the term for *valve acceleration*.

~jM

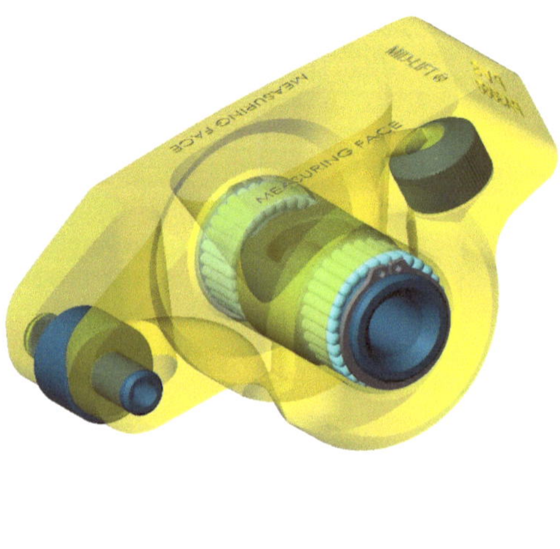

CAM FOLLOWERS

ROLLER, MECHANICAL, HYDRAULIC

Which "TAPPET" is Best?

Each of these could be an article unto themselves, but I am only going to make a couple of observations as food for thought, with a little background to put it in perspective.

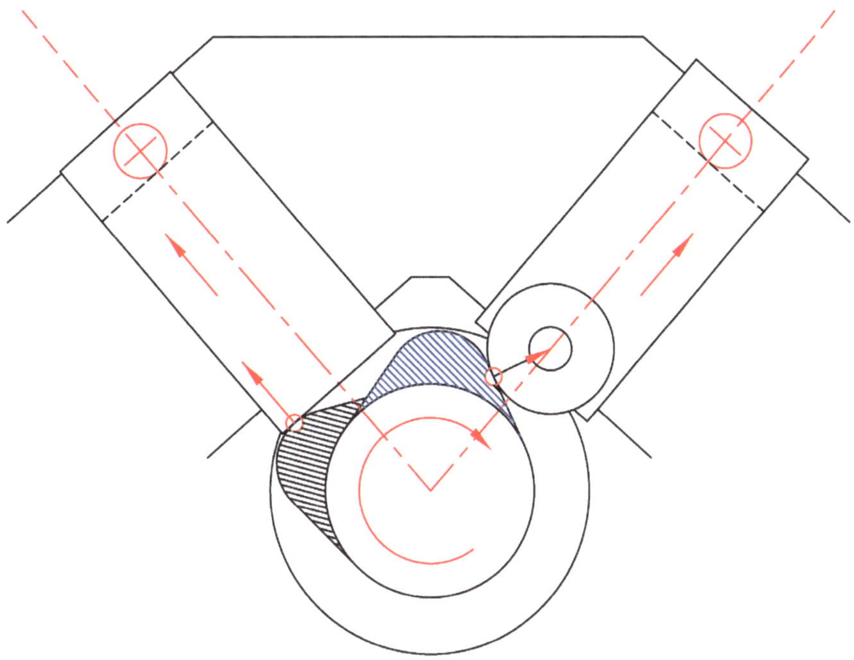

FLAT TAPPET vs. ROLLER

There are two geometries with cams and cam followers of OHV engines (OHC are a third and fourth). These are not the same geometries normally discussed with rocker arms but the word still fits. These two cam geometries are as different as night and day in how they do the same job, which is to use a rotating eccentric to elevate and descend a tappet at prescribed rate of SPEED, DISTANCE and TIME.

The two designs are: (1) the ROLLER TAPPET/CAM, and (2) the FLAT TAPPET/CAM. The following will address only the **CAM geometry** differences between a roller tappet and a flat tappet cam follower; not to be confused by the "hydraulic" variants, which pertain only to lash control and are available in both tappet designs. The Hydraulic cam particulars will be addressed separately.

CAM FOLLOWERS

FLAT TAPPET CAMS 101: You cannot use a roller tappet on a flat tappet cam, nor can you use a flat tappet on a roller cam. They are literally, physically different cam shapes that will destroy the tappets in short order as soon as you try this. In the simplest, purest example, a flat tappet has the cam lobe sweep beneath its bottom to push it up, based on the parameters of lift, duration and velocity that is ground into that cam. This "sweep" along the tappet's bottom surface (which is large radius of roughly 40" that only looks nearly flat), can only take place within the DIAMETER of the tappet's body.

For Chevrolet, this is .843", while the Ford is .875". When the tappet is resting in the closed valve state, the cam's base circle is contacting it in its center. But as the cam rotates, its ramp begins to move outward from the tappet's center as it elevates the tappet upward. If the cam is a race bred high velocity cam, this ramp is "reaching" outward quickly – stretching itself along the tappet's bottom surface much faster as it moves to the outside diameter. The amount of this speed is LIMITED to how far this ramp can reach, which of course is the diameter of the tappet. If the cam ramp was angled too much, too quickly, then it would actually reach beyond the tappet diameter before it raised it up high enough to scoot back under it again to sweep back to the center at full lift; then across center in heading for the CLOSING ramp. If the cam ramp moves outside of the tappet's diameter, it's destroyed. Period.

In other words, flat tappet cams cannot accelerate the ramp any quicker than what the tappet's diameter will allow. This is where MUSHROOM tappets come into play. This broad diameter allows them the width a quicker cam requires. Of course, if you need the extra speed and duration that reaches these limits, and if class rules don't prohibit it, then a ROLLER CAM is what you want to go with. But too often, 90% of the engines built who think they need a roller cam, do not. Besides their higher cost, their greater spring rates rob net torque. In doubt? Compare by turning an engine over with a torque wrench.

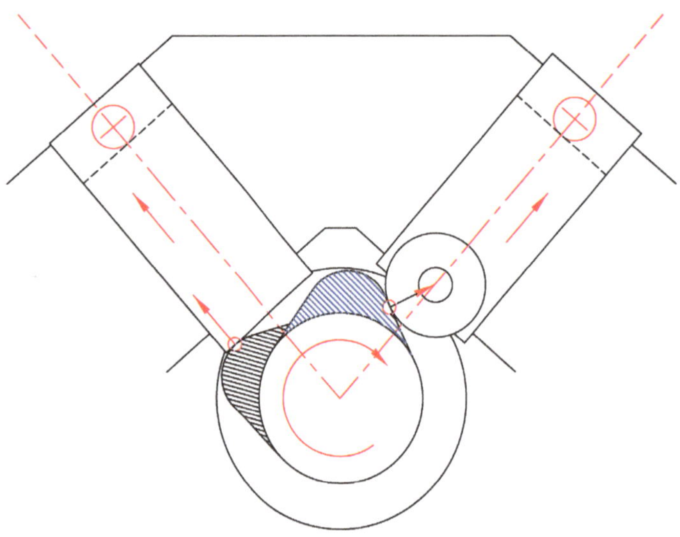

The **LEFT** is a typical flat tappet, while **RIGHT** is a roller tappet.

The two small **red** circles depict the tappet contact points with the cam ramps, while the attached **red** arrows indicate angle of loads. (See previous page for larger detail.)

Note the cam lobe's contact angle to the **roller tappet** axis. *This angle is constantly shifting as the lobe rolls beneath the wheel.*

The **flat tappet's** contact loads are **always in-line** with the tappet axis, as shown here; but of course this sweeps across the bottom of the tappet as the cam lobe passes.

Note: This chapter may contain some repetition of points from other chapters. For continuity most of these have been edited wherever possible, but some have been left intact for emphasis where relevant.

CAM FOLLOWERS

FLAT TAPPET CAMS 102: These are the most “efficient.” They provide the best bang for the buck of all the choices, UNLESS you have one of these very special requirements. Such as a very big cubic inch monster engine, or a big stroke small block with large port heads, or a more nominal cubic inch engine that will be turning 8,500 to 9,300 RPMs. Anything less and you don’t need a roller. Here’s why:

Rollers offer one main advantage over a flat tappet design: They allow the cam engineer to use a very fast velocity ramp for a small base circle. Specifically, raising the tappet faster for less degrees of rotation, something that big inch engines with big port heads can benefit from. You can get a fast velocity flat tappet profile, but it needs two things to make it practical: (1) A reasonable size base circle (not easy in most small block engines, especially if they are stroked – because of hitting the rods), and (2) a large diameter tappet, like at least a Ford, or better yet – a “mushroom” tappet. Both of these things will allow the cam engineer to kick the tappet in the ass with roller tappet velocities.

All cams derive their “lift” from the base circle's radius. There’s no “adding” material to the nose of a cam. Most cams have their peak lift point running only a few thousandths less than the CAM BEARING journal. So when a cam is ground to make it “bigger” (as in more lobe lift), it’s at the expense of removing material from the heal (base circle), which is reduced by whatever that increased lift will be. In some cases this works directly against what is needed for stabilizing the ramp angles to keep velocity “safe.” It also limits the design variables, since the ramps are pulled in to meet the smaller base circle. It's harder for the velocity to be limited to a rate that the springs and parts can safely handle. Because the flat tappet cam's lift geometry is entirely linear to the tappet axis and is limited by the tappet's physical diameter for the range of sweep beneath it, that base circle can only be so small in relation to the lobe's lift, which limits its angle off center and the velocity that angle generates along the ramp.

The **ROLLER** cam does **NOT** have these limitations because it plays by entirely different rules of contact geometry to lift and accelerate the tappet, as I explain on the next page. But those roller cam rules have a price over the flat tappet cam's simplicity, in financial costs, extra spring load drag, and the additional care of the overall package you choose and how you use it.

With all the technology we have available in 2015 and beyond, the pure simplicity and *bang for the buck* performance of a properly chosen flat tappet cam with correctly chosen and installed rocker arms is an unbeatable and attractive combination for making horsepower that lives, and will take your engine safely up through 7,500 RPM, which is where most V8/V6 OHV cylinder heads peak; particularly on engines with stroked cranks and the corresponding higher port velocity that the higher piston velocity generates.

Having said all this... *if money is not an issue,* then you can achieve the overkill assurances that a ROLLER TAPPET will provide you. And if the profile does not need to be crazy fast and huge, you can also hold back on spring rates to minimize the otherwise downside tax of spring rate drag, which I cover in more detail in the subsequent SPRING PRESSURES section. So make yourself happy, get a roller!

HYDRAULIC FLAT TAPPET CAMS: The same advice for the Hydraulic Flat Tappet applies as the best bang for the buck that subject page began with, because it too is a "flat tappet" cam. Many good performance hydraulic flat tappet cams are available, covering all applications from RVs to Super/Stocks and Muscle cars. Although there are a wide range of Hydraulic Flat Tappet cam grinds from the larger cam companies which are capable of 5,800 or 6,000 (they advertise more), *most* choices level off by 5,000 to 5,500 or less, because the industry pushes the roller cam variants instead. But for "101" engine builds where a performance cam is desired, these make a good choice.

CAM FOLLOWERS

ROLLER TAPPET CAMS 101: Unlike the flat tappet operating principles, the cam lobe is contacting the roller wheel on a constantly changing angle of attack to its axis that remains linear in its reciprocating motion. This angle is depicted in the illustrations on pp.85 and 86, where you will notice the red arrows extending from their datum circles at the point of contact between the tappet and the cam follower. These indicate the angle of contact and motion with the roller tappet wheel, to drive the second arrow that runs in-line to the tappet's axis. The only time a roller tappet's contact point with the cam is in a straight line with the cam lobe and pushrod, is when it is lined up with its centerline at full lift. Otherwise, as the illustration shows, the side of the cam lobe is pushing on the side of the roller wheel while it rotates around. This is a very numerically complex geometry which only a computer can solve to determine the precise shape roller tappets require which gives the cam lobe its nearly square shape, often with an inverted dip in flanks of the lobe. As the cam lobe rotates, the lifting values placed on the tappet by contacting the side of the roller wheel are less, so the lobe has to be more vertical (square looking) at this point. When it gets more directly beneath it and the contact angle between the roller's axis and the cam lobe is less, then we see a more rounded NOSE to reflect this.

A term you may have heard before is "inverted flanks." The above is where it comes from, because the cam ramp is actually inverted, or concave, to compensate for this radical contact angle with the tappet's roller axis, particularly on more radical velocity profiles. Sort of letting off steam at these points of lift. There are no physical limits to how fast this cam can be designed for, compared to the flat tappet cam. But the roller tappet is heavier. Also, it takes side loads from the cam as the tappet rotates around the lobe. These are often unnoticed side effects with engines that require a roller cam; but for engines of moderate to reasonable RPMs under 7,500, the hydraulic roller or the flat tappet cam may be better.

ROLLER TAPPET CAMS 102: offer the most freedom in valve lift, velocity and .050" duration specs. But the down side to using these higher velocity cams is that the valve-train's "inertia" requires the increased spring pressure. This is a downer, because if you really didn't need the extra velocity, then the extra spring pressure will significantly increase drag coefficients in the valve-train and rob the engine of horsepower that should be going to the rear wheels. (Moderate-to-high velocity rollers, with reasonable spring rates are the author's favorite choice for the serious engine.)

The NEED for these higher velocities would be from a BIG cubic inch engine with monster ports, turning 6,000 to 7,000+ RPMS, like monster truck engines. Another example would be a big stroke Small Block engine, with HUGE port heads and a huge Manifold to feed it. The last example would be any smaller or standard stroke engine designed for very high RPMs (8,300 and beyond) which have the professionally prepared cylinder heads to use those RPMs. All three of these examples could make use of a fast velocity cam that needed to get the valve open as quickly as possible, known as "area under the curve." To be clear about this, let me clarify that I am speaking about "quick off the seat" profiles. When I say higher velocity, I am referring to launching the valve off the seat quickly, not just fast in the middle range. Huge port engines, with stroked cranks and the resulting fast piston velocities benefit from such cam profiles. But use this type cam in an engine without these requirements, and you have a problem.

A common misconception is the roller tappet is best because it reduces a great deal of friction over a conventional tappet. Not true. Granted, the roller bearings of the roller wheel offer seamless smooth rolling action atop the cam lobe, but the solid tappet has nearly nothing for actual contact against the cam lobe to create a "drag" from friction. The tappet surface is hardened like glass, has an actual radius of more than three feet – giving it the appearance of being "flat" and the cam lobe itself is actually ground on a slight angle (from the front edge to the back edge) to drive the tappet to rotate in its bore, thus keeping wear to nearly nothing.

CAM FOLLOWERS

ROLLER TAPPET CAMS` other virtue is their all steel alloy of both the cam and tappet's roller wheel, allowing for spring pressures that would kill a normal flat bottom tappet. As noted above, the reduced friction benefits of a roller with its needle bearings over a mechanical tappet are negligible. Forget that.

The friction from high spring pressure roller cam engines is quite high. You can feel it and measure it when turning the motor over with a torque wrench (no spark plugs, of course). The rocker geometry is deadly important here, but this is about "cams" so I won't digress on that point too much. The amount of horsepower being used is significant; anywhere from 15 to 40, *or more* with some engines, depending on rocker geometry. If you can turn the engine up to the same RPMs, and if you can reach the cylinder heads` flow capabilities with a flat tappet cam that has enough velocity to work with your heads, you will have at least 20 or 30 horsepower more going to the rear wheels than you would using a roller that unnecessarily uses all this extra weight in the tappets and spring pressure on the valve-train.

You need high spring pressures for FAST ramp cams of racing duration and lift, and you need high pressure springs for the engines that go over 7,600 RPM. Beneath that, you can be very modest about using hi-pressure springs, *especially* if you're using MID-LIFT geometry rocker systems. The in-and-out motion on both sides of the MID-LIFT rockers is ultimate minimum, so the unwanted leverage that occurs on the valve-train from over arcing doesn't exist. The spring pressures are limited to just doing their job. See the MID-LIFT principles described in "Design Geometry." Keep "lash" to .014"–.016".

HYDRAULIC ROLLER CAMS: First came along in 1984 with Ford's introduction of the 5.0 HO engine in the Lincoln Mark VII "LSC," and was soon followed by General Motors' example (I believe first in the Corvette) – and soon the hydraulic roller tappet worked its way into most OHV engines. The OEM mission was NOT more horsepower, but more miles, and second was to reduce warranty costs from traditional hydraulic flat tappet cams. The new hydraulic roller cams for these OEM engines didn't receive any greater "event" timing (hi-performance) than what the traditional hydraulic tappet used, but they did offer near endless miles without excessive wear from poor oil condition. Valve spring pressures were NOT increased because they were "roller."

The Hydraulic Roller soon worked its way into the high performance aftermarket for different reasons, and up to a point offered more longevity and performance – but with a contradiction in packaged hype that continues to this day. This pertains to RPM capability and SPRING PRESSURES.

So, if you NEVER want to set valve lash then your choices are simple, and with today's Hydraulic Roller profiles, these will give you a great result in that same RPM zone, but not any higher. The tried and true Hydraulic Flat Tappet cam is still available and makes a great no-lash performance option with less cost than the Hydraulic Roller version.

~jM

SPRING PRESSURES

There's a myth that you need high valve spring pressures merely because you're running a roller tappet. That's not true. You won't find them on the roller cam OEM engines mentioned above, so the only reason for increased pressure on an aftermarket cam should be for the usual "big cam" reasons. But as with any tappet that is "hydraulic" (roller or otherwise) the RPMs that usually require high valve spring pressures are in most cases beyond the useful reliability of the hydraulic tappet. Controlling the oil pressure "bleed off" of the tappet's internal valve body is promised by some hydraulic tappet manufacturers, but it's not that easy, or realistic. Whether they are called "anti-pump-up" or something else, that automatic valve body that adjusts valve lash for you is now working against you at RPMs over 7,000. Things are happening too fast to allow for a consistent compression of hydraulic oil that must be carefully expelled through a metering hole; especially when that oil is subject to gyrating oil pressures and temperatures; thus: viscosity. Consider 6,500 RPM the safe limit for QUALITY hydraulic tappets; while ho-hum hydraulic tappets will usually bleed down between 5,200 to 5,700 RPM.

There's another myth that you need to run high spring pressures to turn high RPMs. That too is incorrect, depending on what you call high RPMs. Factory Muscle Cars of the 1960s and early 1970s turned low to mid-7000s all day long (once the governor was disconnected), and they had valve spring seat pressures as low as 85 to 95 lbs. Those had both hydraulic and sold tappet cams, from the factory, depending on the car. They didn't easily blow up, either. Simply re-jetting the Holley, recalibrating the distributor curve for full advance by 3,000 RPM, and of course a set of headers, then off to the track on Sunday. Low 12s and high 11s, then back to High School on Monday. A fact.

HYDRAULIC ROLLER CAM

Hydraulic tappets also do not control valve timing and stability as well as solid tappets. If you must have a "hydraulic" tappet for the old fashioned reasons it was created for in the first place, which is avoiding the need to set valve lash, then by all means go with (or stay with) the hydraulic roller tappet. But don't be confused that because there's the word "roller" in its description that it deserves HIGH SPRING PRESSURES or HIGH RPM CAMS. Because it doesn't! If you must have these two things, then by all means decide between SOLID flat bottom tappets and SOLID ROLLER. But do it for the right reasons.

When modifying hydraulic roller factory engines with a new hi-performance cam reasonable spring rates should still be used. Hydraulic tappets don't work well with the upper end spring pressures, so keep your pressures to no more than 135 to 150 lbs on the seat. Open pressures should stay in the low to mid 300s. You can go up in the 375 to 400 range, but open pressures in this amount means you're likely using a valve lift of .600" or more and heading beyond 7,000 RPM. If you're not, then you're throwing free horsepower away with spring rates that aren't necessary. Once you hit 7,000 RPM and beyond, the hydraulic tappet is dead and buried. I don't care what any cam company says. You're best off to switch to a solid flat tappet cam with spring rates no more than 135 lbs seat and 375 lbs open; or a solid roller, with spring rates of 175 to 225 lbs on the seat and 475 to 550 lbs open, depending on your valve lift and cam velocity (how hairy it is). Naturally, these are general yardsticks; but very good ones. The trick is NOT getting talked into something you don't need, can't use, and won't fit – by someone who isn't in charge of all the other components you need to buy and use to make your cam work. Or... to get a cam that works great for "a while," until all of a sudden it has broken something else not covered by warranty.

SPRING PRESSURES

SOLID ROLLER CAM

What are "high spring pressures?" This is relative to the type of cam you're using. If it is a traditional cast iron mechanical flat tappet cam, or a "hydraulic" flat tappet cam, then anything over 135 lbs of seat pressure would be considered high; and could be destructive, or at the very least, short lived.

However, if we are talking about a **billet steel roller cam** with solid roller tappets, then high spring pressures would be anything over 250 lbs on the seat. Although 300 lbs and more is sometimes used – which is very high – considering their "open pressure" may easily reach over 800 lbs, such high spring pressure cams are usually found only on engines of limited operating time, like drag engines for big inch, low weight classes that have monster port heads. However, "*some*" – well built engines for circle track, SCCA and the like, have heads with cams that respond to 300 – 325 lbs of seat pressure, providing the pushrods are .120" heavy wall (or more). Spring maintenance and tight valve lash is the key here. Checking both often for wear is mandatory.

In most cases however, mainstream high performance engines running .550", .600" and even .700" valve lift cams that are turning no more than 7,500 to 7,800 RPM, are too often sold valve spring packages that are **way beyond what is needed** for their cylinder head's flow capabilities, and their reliable use. The excessive pressures are very hard on the valve-train, and without benefit. If the cam is so radical, that it requires 300 or more pounds of seat pressure on an engine that won't go over 8,000 RPM, then something is wrong. There's no "area-under-the-curve" argument to be made that gains power at the expense of common sense and engine longevity. If you can't finish a race, it makes little difference how much "quick power" you've made.

VALVE SPRING HARMONICS

VALVE SPRING PRESSURES are notoriously misused. There are two interesting things to keep in mind about valve springs. One: They are a torsional component; meaning...when they open what is really happening is the individual wire is "twisting" on its own coiled axis as the spring compresses. The second point of interest is that valve spring pressures *after a certain point* do NOT really need to be increased because of MASS, or INERTIA, or RPMs, even though that is what probably 95% of the engine builders mistakenly conclude. Valve spring pressures need to be increased to overcome the extreme waste of valve spring pressure WHEN **HARMONICS** reach their critical state to rob it.

Harmonics come in a series of waves which vary in their degree of intensity. With the first waves of harmonic influence, usually around 7,300 to 7,600 rpm, the end coils begin to dance around, often rotating the spring. The end coils really don't work anyway, they just support the three coils in between. But they begin to leave their mating faces (head and retainer) before anything else starts going to hell.

When the second stage of waves occur, usually 700 to 1,000 rpm beyond the first, the spring now has only about half of its coils compressing, while the rest are going into full static suspension. ***This*** is what the increased spring rate is for: to compensate for the two or three coils that are free floating without applying much (if any) pressure to support the remaining coils and maintain their contact pressure. This increased spring pressure compensates – in a bad way – for this spring rate loss when the end coils are no longer seated and the middle coils are fighting each other. Parts are dancing all over, retainers slap around with their locks, spring shims and cups start floating off their pads and rotating. It's a mess!

SPRING PRESSURES
NON-ROLLER CAM

Harmonics are a direct consequence of many factors which **BAD ROCKER GEOMETRY** exaggerates. But one "constant" element of harmonics which makes them somewhat predictable, is the metal alloy springs are made from. The frequency for steel is different than titanium, which is much higher than steel, ANY steel. The ultimate valve springs are titanium, solely because of their molecular influence with frequency. In fact, I know of several tests that have proven as little as 85 lbs of TITANIUM spring pressure can control RPMs with no valve float, beyond 9,300 rpm on a radical cam small block Chevy. The same engine, with steel springs, went into harmonics and valve float just over 8,500 rpm, and needed over 225 lbs of seat pressure to get there. I make this point to remind engine builders that HYDRAULIC roller cams don't need 150 or 160 lbs, or a flat tappet cam doesn't need more than 135 to 140 lbs of seat pressure on valve-trains that have "proper geometry" and "stable" cam profiles, which will never see more than 7,500 RPM anyway. It takes HORSEPOWER to turn over extra spring pressure. Use only the spring pressure you need to do the job. Keep non-mechanical (hydraulic) roller tappet profiles to no more than 140 lbs of seat pressure, and no more than 385 lbs over the **NOSE**. Mechanical rollers are another animal, which can justly needing more; but they are not necessary in most cases of competition even though the cam companies love to sell solid rollers. (**See: COIL BIND**.)

I hate to say this, but there's been a trend by cam companies over the years to misinform engine builders about SPRING PRESSURE on FLAT tappet profiles that use high tech mechanical (solid) tappets they explain as needed for various, nasty cam profiles. Rather than naming names, pointing fingers, let me explain a couple of things that makes this erroneous (BS).

#1: There is a specific known value of cam/tappet acceleration that can't be exceeded, because of the diameter of the tappet. THIS is the limiting factor on a flat tappet came. Unless "mushroom" tappets are used, this limit won't change. How quickly the cam accelerates is what determines "area-under-the-curve" which is how fast the valve will open to s greater value for each degree of crank rotation. The inertia created by the acceleration rates which are limited to the .842" Chevy and .875" Ford tappet diameters is below, as it has been for **50 years**, the values needed to justify 175 lbs (and more) of seat pressure springs. Titanium springs will shift this limit up, but not by much.

#2: Increased Valve Spring Pressure is used to COMPENSATE for the LOST spring pressure that occurs from HIGH RPM HARMONICS, which are ABOVE 7,800 RPM, and has little to do with substantially controlling valve float on engines running below 7,800 RPM.

#3: The cam companies that are recommending crazy valve spring rates on flat tappet cams (that I personally know of) don't have a clue about the value of correct rocker geometry, and have not tested for it. Therefore, all their data is meaningless. Whatever they say a cam needs for valve spring pressure was determined (if at all) from combinations that had over-arcing rocker geometry which consumed vast amounts of energy going to the wrong places because the rocker arm wasn't doing its job efficiently. Period. Secondly, the cam company making such recommendations is also lacking the data needed on your cylinder heads to know if they need, or can use a fast ramp cam. 90% of the heads sold by the top manufacturers run just fine, and can't take advantage of a truly fast ramp cam that comes close to the "aggressive" version designed in roller profiles; let alone flat tappet profiles.

130 lbs to 145 lbs of seat pressure is all that you need on a well designed flat tappet cam, with correctly installed MID-LIFT rocker geometry (for the least amount of over-arcing and binding). Increasing Valve Spring Pressure to 175 or more pounds is throwing horsepower away, adding unnecessary friction, flexing parts more than necessary and heading your engine for destruction. *Are there flat tappet cams out there that require 165 or 170 lbs of pressure to get the most out of the engine? **YES**. But these are very limited examples where the total combination has been tweaked to the last percent, and **RPMS** are going to exceed 8,300 or more! If you are NOT going to run 8,500 RPM and ABOVE, then considering any flat tappet cam from any company telling you that you need 175, or 200 lbs spring pressure, is a company you should be running away from as fast as you can slam down the phone! ~jM*

ROCKER RATIOS

INCREASED ROCKER RATIO benefits is a question often asked:
"Why do different engines use different ratios?"... "What is the benefit of more ratio?"...and: "What is the consequence of increased ratio?" Here are a few points to consider. In the beginning, as with many engineering ideas pouring in from all over the world, there was no standard. Simple answer. There was no need for one. Rocker ratios are an easy way to increase the needed movement at the valve with less movement at the cam. These were spawned by physical space limitations in engine blocks and obvious benefits to efficiency in multiplying this motion, as engineers realized greater valve lifts over the decades of engine evolution. It was also easier to experiment with valve lift and timing through simpler changes in rocker arms atop the engine, rather than redesigning the block, the cam, and so on as engineers tried increase the engine's power. (This was, by the way, spawned heavily by aircraft engine development during and after WW1.) The 1.50:1 ratio became a good balance many engineers settled on, making slide rule calculations (literally) an easy denominator to deal with. Ford, Chrysler (and others) found a need to adjust this a little more, increasing up to factory OEM ratios in the mid-1.70s. But the reality is, most ratios were chosen for convenience in keeping with traditional values that worked with existing cams, where no real revolution to change was seen.

Keep in mind, that the *effects of valve lift acceleration from a roller tip rocker arm is entirely different from that of a friction pad "shoe tip" rocker arm.* Because of this, much of the aftermarket industry's influence with roller tip rockers on modified Detroit passenger car engines was misinterpreted, even though they would be advertised, sold and used as "stock ratio." (i.e., 1.50:1 for small block Chevy.) Why does this matter?

As an engine builder choosing and buying an assortment of components, often from a variety of manufacturers, you need to understand a little bit about the history of these parts to make sense of the conflicting sales hype you may hear between the different companies which compete with each other. Most cam companies sell rocker arms. Some rocker arm companies sell cams. So of course they each want to sell you their companion parts, but as a savvy engine builder you know that some cam companies don't make the best rocker arms, and vice se versa. Go to a mail-order warehouse and you'll often need your lunch to listen to all the hype. Most important to the above is understanding what stage you are at when changing rocker arm designs (roller tip vs. shoe tip) and the existing NET ratio to your possible increased ratio. LASTLY and MOST IMPORTANTLY is the DESIGN GEOMETRY of the rocker arm you've been using, compared to the brand or change you are about to make.

This applies to both **DESIGN** *and* **INSTALLED** rocker arm geometry. You can choose a 1.50:1 MID-LIFT geometry rocker and easily see as much valve lift per degrees of crank rotation throughout the lift cycle, as possibly using a non-Mid-Lift geometry brand of rocker arm that is advertised as a 1.60:1 ratio.

Here's the short answer to the main question about INCREASING ROCKER RATIO: All else being the same (roller tip for roller tip, shoe tip for shoe tip, etc.), increasing rocker ratio will increase ALL THREE dynamics: VALVE LIFT, VALVE DURATION (note that I did NOT say "cam duration"), and RATE of ACCELERATION (velocity). Bear in mind it will do this on BOTH sides of the cam cycle, OPENING and CLOSING. In "some" cases, this quicker landing of the valve upon the seat from a quicker rocker ratio can increase wear. Of course the most obvious warning is PISTON to VALVE CLEARANCE; although in most cases, "moderate" increases in ratio (i.e., 1.50 to 1.60) won't often create a problem if there was a surplus of PV clearance to begin with. But YOU MUST CHECK to be sure. Always.

ROCKER RATIOS

There's always been a trend to "split ratios" on many engines where a higher 1.60:1 intake rocker ratio may be chosen, then accompanied with a 1.50:1 exhaust rocker ratio. It is not necessary or even beneficial to do this when you are building from the ground up, and not mixing parts you need with parts you have. Wherever stand mounted rocker arms are being used with different valve lifts between the intake and exhaust, the rocker arm's different INSTALLED geometry between the intake and exhaust should be set by changing the height of the valve tips themselves. Since the exhaust valve on most applications will have less valve lift, its rocker arm's pivot point (the shaft) would sit higher in relation to the valve tip. But since the intake and exhaust are likely joined on a common shaft, the only way to change where the shaft sits in relation to the valve tip, is to decrease the valve tip for the valve with the lesser lift; by exactly HALF of the difference in between the two valve lifts. If the Intake valve is .600" lift, and the exhaust is .500" lift, then the exhaust valve tip would be .050" LOWER than the intake. This has the same effect of raising the shaft in relation to the valve tip. Of course, you have to plan for this in advance. It can't be a last minute change. On individually mounted stands, such as the big block Chevy, this doesn't matter. You can match all valve tips to the same height, and designate your stand height for each, as needed to achieve MID-LIFT geometry. Rocker ratios should remain the same, for several simplicity reasons to design and use, and the differing specs needed between the intake and exhaust valve should be ground specifically into the camshaft. Don't use the rocker arm for a "tuning" tool.

THE REAL BENEFIT to increasing rocker ratio, is that it allows for a larger base circle camshaft, which on a full competition engine, is often a good benefit. The rotational torque, flex and "whip" imposed on a camshaft operating with high spring pressures at high RPMs is horrendous. Increasing the cam's base circle aids the cam engineer in adding precious material where it resists this, and also in the cam design itself, there is more room for creating a "fast" ramp (more velocity). There is a balance to this though. The rocker arm must be designed to withstand the additional beam loading without adding more spring board effect to the rebounding that this extra pressure induces. So too must the pushrods, tappets and all mounting hardware be adequately beefed up to accommodate the extra pressure that is proportionately increased with the percentage of increased ratio.

It should be noted that only in flat tappet camshafts does this increased ratio give the most relief to the cam's base circle. But because a flat tappet cam is a "friction" component, rubbing directly atop the cam lobe, it also suffers the greatest limitations to increased valve spring pressures. Roller tappet profiles have no real mechanical limits to how fast the tappet (and valve) can be accelerated, as that imposed by a flat tappet profile. But of course, there ARE limits to inertia, mass (weight) and velocity which are cut in stone with specific material designations and designs. For your information, we've designed and tested (successfully), as have other rocker arm manufacturers, ratios from standard 1.50:1 to over 2.20:1 for NASCAR designated engines. But the cams were specifically and carefully designed for these. When all was said and done, everyone came back to earth and still relied on designs between 1.70 and 1.90. Even though some may have found one lowly horsepower here or there stepping outside of this, there was always a counter-benefit that offset the gain; usually it was longevity.

One last point, "historically" the largest inventories of master cams with the best cam companies are based on their original COMPUTER GENERATED profiles being for a **1.50:1** small block Chevy rocker ratio, REGARDLESS what engine application and rocker ratio was used when the manufactured cam for a customer was generated. The cam company simple adjusted the *predicted* gross valve lift and duration on the cam card for whatever the rocker ratio change was for whatever engine that particular MASTER CAM grind was applied to. CNC Cams have changed this; but MANY cams still sold for a Big Block Ford were "developed" in a computer from the DNA of a *small block Chevy!* (So much for **ratios.**)

ROCKER RATIOS

RATIO DIFFERENCES: From time to time I was asked why I don't make rockers that duplicated the same ratios as factory applications, like the 1.73 ratio Ford, instead of only offering the 1.70. The short answer is the market didn't demand it. But, even though it overlaps some of the points made above, my long answer is as follows:

1. "Ratio" is a "snap-shot" only at full valve lift that does not account for the accurate ratio of any rocker arm throughout its arcing cycle; which is a different dynamic that affects cam performance (at the valve) drastically, and is known as "area-under-the-curve." Also, because "geometry" of SHOE rockers is entirely different in the way the rocker arm responds to SPEED and DURATION at given points in between CLOSED valve and OPEN valve – when RATIO is determined against roller tip rocker arms. So we don't need to "nit-pick" with our rockers in trying to duplicate arbitrary ratios the OEM factories came up with.

2. You will also see that there is NO STANDARD between one company's design and another, because before MID-LIFT was developed, all companies chose random CLOSED valve designs with VALVE lengths and rocker arms fixed in the CLOSED valve position with NO REGARD for the wide range of effective ratio changes the rocker arm passes through BETWEEN CLOSED valve and OPEN valve. I am not criticizing these companies. I am pointing out *why* this wide range of designs existed. Because as an engine builder, if you buy rocker "A" with a 1.60:1 ratio, build an engine, race the engine, then break a rocker arm; it would be nice to know that if you changed to rocker "B" of the supposedly same 1.60:1 ratio that your piston to valve clearance and overall performance remained the same. Guess what? It likely will NOT. Why?...

3. If you check different companies you will see their NET (real) ratio, including Ford, Chevrolet and GM, are not exactly accurate to what is "published" or specified as the "design" ratio. None of these companies, ever specified a standard of any accuracy, nor did they educate anyone as to what their ideal settings of rocker heights for GEOMETRY was, until *after* MID-LIFT was published; then studied; then adapted by them. Now it's the only geometry that Chevrolet or Chrysler recommend. *What difference does this make?* If you change the PIVOT height of any rocker arm when checking ratios, you'll notice that the NET RATIO CHANGED also. There is no exact value for what ratio will be, WITHOUT a known geometry value.

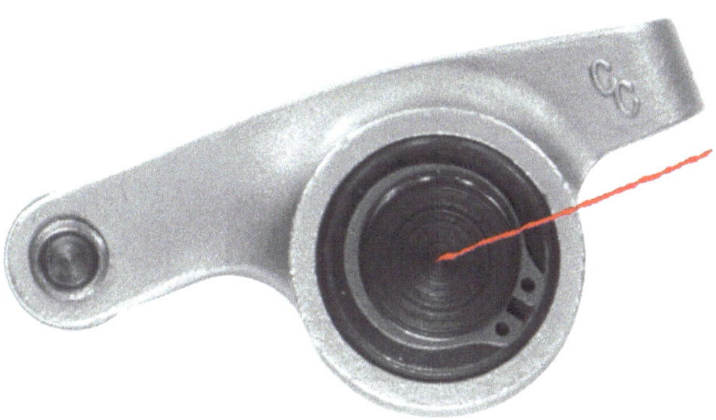

The roller axis is shown horizontal to the trunnion, while the **RED LINE** on this popular brand rocker illustrates where the pushrod cup *should be.*

This .150" too high cup has already over-arced to make it impossible to control geometry or lock in an accurate **ratio**; which although is fixed, will shift around depending on pushrod lengths (installed geometry).

Because it is a stud mount rocker there is no way to correct this. It represents the error of many bad geometry brands.

ROCKER RATIOS

4. The ratios were often selected NOT for any engineering reason, but to "match parts." There were no engineers who found a secret to anything in the valve-train that established a "1.73" ratio was better than "1.70 or 1.75." Usually oddball ratios are derived by a stack of linear, compounded measurements of all the components from the cam centerline up through the tappet, then the cylinder head mounting points for the rocker studs, to the valve centerlines... *and*... while trying to make as many of these parts fit EXISTING and FUTURE engines. Hence, the same rocker arm was used for BOTH the 351 Cleveland and the 460 Ford, while the ROCKER and entire VALVE-TRAIN GEOMETRY between each of these engines is as different as night and day!

If the above chaos for using a critical component on two entirely different engines wasn't enough, the DESIGN Geometry ("DG") between these two is on the moon. The 351 Cleveland has an 8.0° Intake rocker DG, and a 13.5° Exhaust DG; while the 460 Ford has an 18.0° Intake and a 23.0° Exhaust DG. So if you are using the same rocker part number on *both* engines, that means the two extremes have a 15.5° spread (351 Intake vs. 460 Exhaust). The rocker arm only rotates approximately 22°, and no matter how you install one or the other, it is going to start life being 3/4 off angle from its total rotation. Now that's cheap!

If you were to measure the amount of CRANK rotation that it takes to open the valve with any engine using properly installed MID-LIFT rocker arm geometry, you would find that MORE VALVE LIFT occurs THROUGHOUT the valve lift cycle from LESS CRANK rotation than any other rocker arm design, especially factory SHOE tip designs, which have an entirely different tracking velocity than a roller tip rocker, regardless of the manufacturer.

~jM

The Patent Pending PRO-SHAFT™ system is unique in that each rocker arm was lengthened as needed to be perfectly aligned with the tappet bores and center the roller on the valve tip of this Ford 351 Yates NASCAR head. The factory's in-line tappet bores forced the OEM pushrod geometry to lay way over, making the rocker arm geometry twisted and way off. This solved that.

CHOOSING VALVE LIFT

Engine builders are too often told what the airflow of a given cylinder head is while reciting where the peak flow occurs, and so cam and valve lift is chosen based on this ultimate point where there is no increase in flow for any additional lift of the valve. This is WRONG. What you want to think about is the following:

1. Look closely at the flow graph of both INTAKE and EXHAUST ports, as measured in at least 0.050" increments. (0.100" is too big a jump.)

2. On the intake port, calculate and note the *percentage* of change from each 0.050" increase in valve lift. This will be significant through the middle valve lift range, and start slowing down with less increase as the port nears its peak.

3. When the port "doesn't make sense" anymore, by giving you only 2% or 3% increase in air flow when you increased the valve lift 8%, or 10%, or 15% from the previous measurement, then the port is not being efficient in proportions to the wear and tear you are adding to the valve-train. You'll see 15% and 20% and more, jumps between measurements, until the last .100" of "usable" valve lift. Then you will only see 6% or 8%, which is when the port is just about dead, then you'll be down to only 1% or 2% or so. And the port died about 0.050" before then! This is when you want to pull back before you send the valve higher than it needs to go. When the port has already completed 80% of its work, you're only getting 2% gain of the REMAINING 20%, not the entire valve lift range. The stress exerted on the springs and other valve-train components for this paltry amount of flow that the engine never can use is not worth it. It doesn't make sense to open the valve another 10%, to only get 2%, when all your gains was double-digit for each step before this point.

Will the head use another 0.050" valve lift to give you 2% more air flow? Yes, *but the engine won't. Here's why.* When then INTAKE valve is really working, 80% to 90% of its cylinder filling to the "engine" comes from the valve lift between roughly .200" (less valve lift still has the piston rising) and 80% to 90% of its peak valve lift, as defined by when the port efficiency quits working. This point is when you have nearly nothing for percentage of increase in CFM. The port has stopped delivering these powerful and "responsive" flow movements in proportion to the valve lift. The benefits are nearly nothing compared to the increased stress to the valve train components. That last 0.050" you may have added, to only get 2% more flow, when 90% of the port work has already been done, may very well have pushed you into the red zone on your spring's coil bind clearance, which decreases the RPM limit before the spring goes into the first wave of harmonics. You always want a surplus of coil bind; more than 0.050". Set coil bind clearance for a minimum of .100", and the springs will love you for it.

With the EXHAUST valve it's even more important, because 80% of its work is done in the first 40–50% of its lift, beginning (unlike the intake valve) instantly upon opening. You want to use the same making sense formula. When the last 0.050" increase in valve lift is only giving you 2% or 3% increase in flow, DON'T go to that last .050". Stop the valve lift short of it, and call that "peak flow," because it is!

CHOOSING VALVE LIFT

The foregoing pertains to full and moderate competition. When you have a "pro stock" level cylinder head on the flow bench, you will see double digit gains way up in the high valve lift range, .800"+ There are formulas between VALVE SIZE and PORT SIZE that determines the percentage of gains per each station of lift. However, the above should be used to dispel the incorrect use of too much valve lift on "out of the box" CNC or aftermarket castings which have these advertised flow figures that mistakenly imply the greatest valve lift figure they use is where your valve lift should be. IT ISN'T! Don't trust anyone's advertised anything. Pay someone you trust to flow the heads for you. Just because a port has been CNC machined, or even hand ported, doesn't mean that the cross sectional area from beginning to end matches the formulas needed for that valve diameter and valve bowl design. *The truly optimized "laminar flow" head that creates double digit flow jumps all the way up the ladder without trading off low lift flow (and the velocity that goes with it) is VERY expensive pro only stuff. Anyone can take a hand grinder and make a sewer pipe out of heads, but the real test in their quality is the smooth gains in flow from increment to increment all the way through the cross sectional area values that the valve diameter dictates. When the flow prematurely flattens out below this valve area value, to less than 10% increase from the previous 0.050", the port has reached its limit. STOP the valve lift here.*

NOTE: See *CHOOSING A CAM* (next chapter) for insight on selecting the DURATION and LOBE SEPARATION you need for your particular engine combination.

CUBIC FEET PER MINUTE @ 1.5" MERCURY

Baseline tests of cylinder heads intake port ONLY-Radiused entry

VALVE LIFT

Thousandths inch	T-Bird-Small valves	T-Bird-SCJ valves	CJ/SCJ head
.200	104.7cfm	131.1cfm	115.2cfm
.300	163.2	177.0	160.8
.400	194.1	206.1	200.4
.500	215.4	220.5	229.5
.600	229.5	229.5	254.4
.700	238.2	239.4	276.0
.800	240.6	240.6	291.6
	EXHAUST PORT FLOW TESTS		
.100	37.6 cfm	SAME RESULTS	40.4 cfm
.200	74.6	as CJ/SCJ	74.8
.300	100.8	HEAD	101.0
.400	111.2		116.6
.500	116.2		127.2
.600	117.2		131.2
.700	117.4		132.6
.800	No Test		133.6

Excerpt from a 1982 Super Ford article by the author, the flow figures to the left compare a Ford 460 T-Bird head with the 429 CJ/SCJ.

The first column represents the smaller stock T-Bird; the second is modified with CJ valves; and the third column is the larger port 429 CJ/SCJ head. (The top group is Intake and the lower group is exhaust.)

As with advertised flow figures of heads sold today, this chart depicts the flow figures in .100" steps. To use the .050" valve lift formula mentioned above, simply split the .100" values and apply in between.

Even with theses .100" steps a trained eye can see the first column for the "T-Bird" size valve stops flowing BEFORE .600" valve lift. It moves from 215 CFM at .500" VL, to 229 at .600". It's only gained 14 CFM for .100" VL, which is only 6% gain. When using .050" VL formula, this is only 3% gain from .500" to .550". It would be okay to use a NET Valve Lift (VL) up to .600", but in all reality .550" would be fine.

A close look at the EXHAUST Valve column beneath it is even more sad. The subject of this article focused on the grossly inadequate, and poor design of the "dog leg" exhaust port, and that result is shown here, where this port is dead and buried at .500" VL, having only moved to 116 CFM from 111 at .400" VL. Applying the .050" VL step means this is another less than 3% gain at .450" VL. *~jM*

CHOOSING A CAM

The following are some basic fundamentals relevant to selecting a cam. It is by no means a cut in stone blueprint where some of these variables can't be stretched. But this should keep your feet on the ground to avoid losing perspective of what is important, over what someone is hyping to make a sale, or worse, failing to give you what you really need.

> **NOTE:** There are THREE components of cam design which must each be chosen for your engine. These are: **VALVE LIFT, CAM DURATION, and CAM LOBE SEPARATION**. When speaking directly to the cam company of your choice for a custom profile, either you or they will translate a fourth important value of the equation into "cam talk" and that is the **Rocker Ratio**. Make sure the cam company and you speak in terms of "**Lobe Lift**" on the *cam* itself to avoid a changed rocker ratio from being missed in the discussion. *All discussions aim for "net" valve lift.*

CHOOSING VALVE LIFT requires knowing the REAL value of your cylinder heads airflow, and in particular – exactly when that value comes to its end – to avoid a mistake engine builders often make of choosing a cam with more NET valve lift than they actually need. Hopefully, you've read the preceding chapter; but if not, this would be a good time to do it.

DURATION: As noted in Terms & Definitions is a value that is spoke of in a couple different ways, usually Gross and "050" (O-Fifty). It is the .050" tappet lift that you want to focus on, without paying much (or any) attention to the gross duration specs; or any other value of tappet lift (i.e., .020") for that matter. O-fifty is the standard everyone has settled on for the last 30-plus years, for good reason.

Here's the fundamental point with DURATION: *The higher the RPMs, the longer the valves need to be held open at a given lift to accommodate giving the head time to respond with additional airflow.* At lower engine speeds, having the valves stay open longer than they need (for that RPM) will show up with sluggish response. So we are making choices that are always a compromise aimed to enhance the major RPM curve being used.

Duration is relevant to TWO values: CUBIC INCHES of your engine, and RPM to be used. To confuse you even more, both of them offset each other; and not taking that into consideration can get you into trouble, or at the very least – leave you short of what you require in this department. Without making a big deal out of this, let me just repeat a small point for comparison: If you have a large cubic inch engine for the "family" size it comes from (i.e., a 430 CI Small Block Chevy, who's "family" is 302–350 CI), then this engine has larger demands for cylinder head requirements (port size) than it was originally designed. But **most important**, is this increased size obviously came from an **INCREASED STROKE**.

Increased stroke engines will have the piston operating at a faster speed, "per RPM," than what this engine originally had its components selected for. It "thinks" it is running 7,000 RPM when the tach only says 6,500. Accordingly most "stroker" engines don't turn as many RPMs as their smaller counterparts. However, they need the same duration cam. So with stroker engines, don't let the fact that you know you'll be turning less RPM detour you from a duration that a smaller engine turning more RPM would require. A Big Block engine was designed from a family of larger cubic inches to begin with, so the term stroker for this family usually denotes 500 cubic inches and up, topping 700 cubic inches.

Note: This chapter may contain some repetition of points from other chapters. For continuity most of these have been edited wherever possible, but some have been left intact for emphasis where relevant.

CHOOSING A CAM

Here are some general rules based on **.050" tappet lift DURATION**. Your cam company's recommendations may differ; if so, ask they for specifics on why.

> **NOTE (1):** The following pertains to **NATURALLY ASPIRATED ONLY**, and predominantly WEDGE chambers. (Supercharged and Turbocharged engines would have a higher emphasis on the exhaust.) HEMI combustion chambers (i.e., Chrysler or Ford BOSS) would require **4° to 8° more** than wedge.
>
> **NOTE (2):** The following duration specs pertain **ONLY** to the **INTAKE VALVE**. You would automatically add about **4° to 8° more** for the **Exhaust**, for most maximum performance cases.
>
> **NOTE (3):** The following is for the original, stock short-block "family" of engine size designs. If you have STROKED the OEM engine, AND you are going into similar peak RPM limits, then you should consider ADDING **6° to 10° more** than what is recommended below, to allow for the increased piston velocity the stroked crank imposes on the cylinder heads.
>
> **NOTE (4):** The RECOMMENDATIONS below are for rocker systems set up for **MID-LIFT** geometry, where both sides of the rocker arm adhere to the original MID-LIFT specifications to achieve the least amount of wasted motion on each side of the rocker arm.
>
> **NOTE (5):** If you're using rocker geometry that is not designed and installed for Mid-Lift, then add an additional 5° to 10° degrees MORE duration may yield better performance to compensate for the lost cam information *at the valve* due to increased motion of the push-rod and roller tip sweep.

For engines that will not see more than **4,500 rpm** (basically stock, but warmed over with headers, good intake and adjusted fuel curve), 210°–220°.

For engines that will not see more than **5,500 rpm**, with the above modifications and still basically stock short block specs: 220°–230°.

For engines that will not see more than **6,500 rpm**, with the above modifications and still basically stock short block specs: 230°–240°.

For engines that will not see more than **7,500 rpm**, but have more serious cylinder head capability with good exhaust, intake and proper fuel curve but basically stock short block specs: 240°–250°. NOTE: If the engine was to be predominantly operated in the lower ranges of this RPM limit, or needed good low end torque, I'd recommend advancing the PHASING of the cam with the crank by 4°. (Degree it in to advance it this amount.)

For engines that will not see more than **7,500 rpm**, but have PORTED cylinder heads with TUNED HEADER exhaust, matched intake and proper fuel curve with matching race prepared short block of increased compression and adjusted lobe separation to accommodate: 250°–260° in a "straight up" phasing of cam, OR: 260°–270° with advancing the PHASING of the cam with the crank by 4°. (Author's preference.)

For engines that will not see more than **8,500 rpm**, but have serious PORTED cylinder head capability with TUNED HEADER exhaust, matched intake and proper fuel curve with matching race prepared short block of increased compression and adjusted lobe separation to accommodate: 260°–270° in a "straight up" phasing of cam, OR: 270°–280° with advancing the PHASING of the cam with the crank by 4°. (Author's preference.)

CHOOSING A CAM

For engines that *WILL SEE* ***more*** than **8,500 rpm**, and have PROFESSIONALLY PREPPED cylinder heads with well tuned equal length OPEN HEADER exhaust, high flow intake ports and generated equal length runners; proper fuel curve with premium race prepared short block of maximum compression and adjusted lobe separation to accommodate: 285°–295° in a "straight up" phasing of cam, OR: 290°–305° with advancing the PHASING of the cam with the crank by 4°. (Author's preference.)

LOBE SEPARATION: Is – like many engine terms – used to describe one aspect that is actually doing something else. As mentioned on p.121 for this term, this determines **CAM OVERLAP** (the period when both intake and exhaust valves are open simultaneously). This overlap is what determines *Net Effective Cylinder Pressure* (**NECP**), or the actual running compression your engine sees, as measured in pressure. This is different than the theoretical pressure attained from your **STATIC COMPRESSION RATIO**, which is derived from dividing the net volume of your combustion chamber by the total volume of your cylinder, as measured from perfect Top Dead Center (**TDC**) and Bottom Dead Center (**BDC**) of the piston. This has no real value for determining NECP because the cam's overlap is the final dictator of that. *The more cam overlap you have, the less running compression you have.* High RPM engines need more cam overlap where the valves are open together for a longer time to "purge" the wasted gases with the in flowing intake charge. But the price for this longer overlap is that we LOSE effective NET compression (NECP), or what I call the "running compression." To recover this loss we increase the STATIC compression ratio. This order of priorities is important, because changing compression ratio without understanding the consequences can make for a hot running engine. The real blame when this happens, is often the confusion of terminologies that even the most seasoned professionals of cam companies, cylinder head manufacturers, and top shops can make, and too often do.

> **NOTE:** Here's the fundamental RULE with LOBE SEPARATION. *The higher your STATIC COMPRESSION RATIO, the LOWER your LOBE SEPARATION.* The range of lobe separation most often seen is between 104° and 114°. As noted in the definition for lobe separation, this is the actual physical angle between the cam lobe centerlines of the camshaft. Not to be confused with crank degrees, as most all other cam information is in reference to.
>
> So, if you want to skip past all this background information here's a yardstick of REAL numbers: For a naturally aspirated engine, which is what all these recommendations are aimed at, and you have a WEDGE style cylinder head (i.e., SB or BB Chevy or Ford) with 8.5:1 to 9.5:1 CR (compression ratio), your cam should be around 112°–114°. If you're CR is up around 14:1, it should be around 107° to 108°. If you're more than that, then your cam's Lobe Separation could be best at 106°.
>
> If you have a **HEMI COMBUSTION CHAMBER**, then all of the above **lobe separation** figures will *drop* by 2° to 6°. The highest Lobe Separation value you'd want on a Hemi (Chrysler or Ford BOSS) is 110° and the lowest you'd have is 102°, depending on your compression ratio. A **13.5:1** CR **likes 104°** Lobe Separation on most American designed OHV Hemi RACE engines; while lower compression ratios would prefer 108° to 110°. Why? Because Hemi combustion chambers are more efficient than wedge cylinder heads for their valve-to-valve cross-flow, thanks to the inclined valve angles each leans into the other with. However, their more uniform "hemi" shape (compared to that of a wedge chamber) is actually *slower* in its efficiency of burning the air/fuel mixture, because there is *less* turbulence. Accordingly, the hemi requires more time to purge the cylinder, and more cam *duration*. But the intake valve's closing point needs to be shifted 4° with the piston, by comparison to a wedge, to compensate. Wedge engines are notorious for faster burning flame propagation thanks to their turbulence.

CHOOSING A CAM

Any compression ratios you have that fall in between these outside examples should be adjusted evenly between their spread. For instance, if you have 11.5:1 to 12:0:1 CR, then 108° to 110° would be a safe choice.

An additional dynamic that should be considered in the above, as well as with duration choices, is that cam ramps have varying velocities from one company's design to another, and for one proposed mission versus another. This difference is NOT quantified in the cam company's CAM CARD. A cam with a quicker ramp (more velocity) is in essence a "bigger" cam than a like duration version which is more moderate on the acceleration rate. In such cases, you would treat the chosen lobe separation and duration accordingly from this insight. In other words if a cam is very aggressive, then the lobe separation might need to be tweaked to a 1° or 2° greater number because the overlap is greater at the middle of the cam ramp area, which is a reference point the cam card and you are not measuring from. Your cam manufacturer should be able to fine tune their recommendations for you when you are getting this nitty-gritty, but you won't get such answers from the usual "call center" kid answering phones and reading from an advertisement flier. You need to speak to the cam engineer, or one of their sharp assistants who are up on these differences. To CHECK that you're getting the right info, or just to know what you're comparing, you need one of the cam checking gauges and software, available from several companies. Andrews Products and Audie Technology are two such suppliers. Once you understand how important this is from having done it, you'll never look at your engine building choices the same way again. If you have access to an engine Dyno, then this should be standard operating procedure.

Remember, when you're talking about cam *lobe* designs and comparisons, you're doing it from a fixed point on the cam lobe, which is the point where the tappet has been lifted from the base circle by .050" of an inch on the opening side, until it returns to .050" of an inch on the closing side. You can have TWO cam lobes of identical durations, measured at the same two points, while even having the same CAM LIFT; but each can have an entirely different event velocity in between these points; and thus, an entirely different affect on the engine.

Most of the established cam companies have several grinds with nearly identical lobe lift and .050" duration, but of different velocity events. So unless you have a cam measuring tool to trace and quantify profiles for yourself, you need to trust your cam company/engineer. But be sure it is the cam engineer who understands this, or a very savvy salesmen with his finger on the pulse of their products. Not someone sitting in a cubical who's unfamiliar with the nitty-gritty of their various cam models. The right guy will know the dyno results and track history where one model cam of similar specs was better for a specific application over the other. Once you have enough experience with specific grinds, it will be you that the cam company is asking for technical details. From your feedback and others who've paid their dues, they can tweak their variations on that model and help other engine builders dial in a profile for their specific need.

~jM

SETTING VALVE LASH

STUD & STAND MOUNT ROCKER SYSTEMS

BY JIM MILLER

Valve Lash: Setting it right, once and for all

Valve lash is one of the most important aspects of engine tuning, and yet it is taken for granted more often than not. Valve lash can "direct" a sharp engine builder as to the "next" version he or she may want to try in an experimental cam design, simply by varying the lash around a little, and staggering the lash in various ways between the intake and exhaust. But first, valve lash needs to be set *properly*, and it needs to *stay* set. Like anything else, there is a technique to setting valve lash. The following, although simple in principle, is an outline of that principle. Done right, your lash will stay set for a long time, with minimal change (if it does change much; you have other problems).

Most high-performance cams and valvetrains are going to have an adjuster which is going to have a lock nut—whether we are talking about "Poly-Locks" on stud-mounted rocker arm systems, or whether we are talking about adjusting screws and lock nuts on a shaft-mounted rocker system. The following is related to the two-piece locking adjustment; standard self-locking (interference fit) adjusting nuts are not applicable to this instruction, nor recommended for performance usage.

Whether we are talking about the poly-lock adjusting nut or the adjusting screw with accompanying lock nut, the outside nut is what must be used for tightening (torquing) down to final specs. The center Allen screw for the poly-lock is suppose to be the locking device, but it is not. Let's take the poly-lock first.

On the poly-lock the outside nut is used for adjustment, while the inside Allen set-screw is used to "bottom-out" on the top of the rocker arms mounting stud. You normally set the outer nut with one hand while you hold a feeler gauge in the other, thus setting the lash. Once set, you would normally turn the Allen set-screw until it was "tightened." *Do not set the lash this way. Rather:*

This shows the lock screw (left) that takes the Allen wrench/socket and the adjusting hex nut (shown at right) that is used for both the lash adjustment and tightening of final torque during lash setting. On rocker arms with adjusting screws (first photo), the outer nut is still used for torquing to final tightness. However, the Allen wrench needs to be used for the initial lash setting.

Here is the initial setting of the valve lash before any serious tightening has been done. Note the two o'clock/12 o'clock position of the Allen socket (ratchet) and open-end wrench.

The wrenches are now backed off from the lash "clock" position, approximately "two hours" of rotation. Your particular feel may require adjusting a little more or less than this after some trial and error.

The Allen wrench is now tightened back up to take up slack. Be sure not to allow the boxed-end wrench to move while tightening the Allen screw, since this is the lash adjustment that will be tightened back next.

The critical "two-handed" torque is used to lock the little Allen set-screw. Without moving the outer adjusting nut, turn both to the original "two-o'clock" position.

POPULAR HOT RODDING/47

EDITOR'S NOTE: *Author's article in the* June 1989 issue of *Popular Hot Rodding laid out the following...*

For decades good engine builders have blamed other things in trying to figure out why their valve lash settings wouldn't stay locked down. Often they break set screws over tightening them, or they aren't set tight enough and lash won't stay set. The 7/16-20 high tensile fasteners can easily withstand 40 to 45 ft/lbs of torque, provided they are tightened down properly. The trick is that you do NOT tighten the *set screw*.

Rocker arms using an external lock nut on the adjusting screw (Stand Mounted Systems), or rockers that have Allen set screw inside the adjusting nut (Stud Mounted), can each benefit from the little known technique called **PRECEDING**.

This technique will assure your adjusters stay locked down, and it will avoid breaking the set screw in the process. The following assumes a STUD mount rocker with adjusting nut. But regardless which rocker system you have, the tightening principle is the same.

FIRST: With the set screw loose, use a boxed end wrench on the adjusting nut for stud mount rockers (or the Allen wrench used on the adjusting screw of the STAND mounted rockers), to set the valve lash as you normally would, being careful to take note of *where* the wrench points when you have the setting you want, referring to it as if it was an hour hand on a clock.

SECOND: Now rotate the same wrench COUNTER-CLOCKWISE 1/4 to 1/3 of a turn and stop. (This will *temporarily* increase your lash setting.) Being careful not to move your lash nut, now turn the SET SCREW CLOCKWISE to take up the free-play until it bottoms out, remembering that your *setting position is 1/4 or 1/3 of a turn loose.* (On STAND Mounted rockers you'll be turning the outer LOCK NUT clockwise to take up this free-play.)

THIRD: Use **BOTH HANDS** to hold the Allen wrench *and* the boxed-end wrench to turn the set screw and the adjusting nut (or adjusting screw and outer nut) simultaneously, tightening both to the same hour position your wrench was at when the valve lash was set in the first step. In this final step you are applying the locking torque to BOTH the adjuster and the set screw (or the outer lock nut and adjusting screw if it's a STAND Mount system) so they preload within themselves EVENLY, which avoids unnecessary pressure to either one individually.

> **NOTE:** As you most likely construed, the reason for the varying preload in the second step of backing off between 1/4 and 1/3 of a turn is because each mechanic has a different "feel." So you will need to experiment to precisely stop on the exact spot which preloads back to the desired lash setting you measured in the first step when there was no preload. After two or three tries you'll get it and the remaining rockers will set quick and easy. (Avoid 3/8-24 adjusters if possible.)
>
> ~jM

INDUCTION SYSTEMS

The induction system of an engine includes the carburetor(s) or fuel injection throttle body, intake manifold, intake ports of the cylinder heads, and the intake valve. Any peripheral accessories such as "nitrous" injectors can be argued as part of the induction system, but for the purposes of airflow management and its influence on camshaft choices, we are going to omit the variables which could be attached to this topic and dwell mainly on "naturally aspirated" (non-supercharged) engines.

Too many times aspiring engine builders get confused in the maze of terms used to describe building engines. That was one reason I created *Terms & Definitions* for the MID-LIFT.com web site in 2003. Once you understand a few principles of the engine's operation it becomes a little clearer where the attention for performance needs to be focused. Cylinder heads and manifold modifications are one of these key systems which needs to be understood before moving on to the myriad of other factors which affect this system, like the valve-train. To put this in perspective I like to refer to an analogy which everyone has seen but perhaps paid little attention to: The heating system of a warehouse.

The air ducts running along the ceiling or wall represent the basic laws of volume and physics which the induction system of an engine also adheres too. These can be either rectangular shaped ducts, or round ducts; just like the ports of an engine. But let's envision a round duct system for example. The main duct leaving the central heater is large, but as it continues down to the end of the warehouse it keeps reducing in size ever twenty feet or so. Not so coincidentally, there are output vents every ten or twenty feet. The main feeder duct requires that its diameter reduces in "area" by a similar volume as the vents which allow these fixed amounts of blown air to be distributed. Finally at the end of the main duct, it has reached its smallest diameter (or rectangle) and the last vent still allows forced air out, at the same rate and velocity as those output vents near the main heater, where the volume is higher, because it has maintained a constant pressure as determined the volume of air moving through the system. PRESSURE and VOLUME. But there are two forces at work: VOLUME and VELOCITY. The velocity is driven by the pressure, but determined by passage area size, as measured in square inches.

In the induction system of an engine, we sort of have the reverse perspective of our heating system example. Instead of a blower fan located in a central heater forcing air down diminishing sized ducts, we have eight different *ports* (or six, or four, depending on the engine) "sucking" through a centrally merged duct (carburetor or plenum), but the principles are the same. I should clarify at this point that the individual cylinder head port and manifold runner every given cylinder has its own rules of dimensions too, which I will address in a moment. But for now, I want to draw the comparison of "system" requirements for the whole engine; and this will focus around a carbureted V-8.

ONE of the EXTRA denominators separating our heating system example from an actual racing engine (which also separates "good" manifolds from "great" manifolds), is the fact that in heating ducts we are dealing with a constant pressure, under one direction of flow. However, in a racing engine we are dealing with a dynamic range of pressures and two directions of flow with separate and dynamic forces driving each, with two directions of "flow." Another dynamic are the different effects of "wet-flow" verses "dry-flow," but we'll stay away from that element of influence to stay focused on the airflow basics.

The first thing I want to emphasize is that the velocity within the port must be controlled. The speed of the air in an engine is a result of the "sucking" force of the piston – and the engine's revolutions per minute (RPM) in relation to the valve timing which controls when that piston's sucking force begins and ends. This is dictated by the camshaft. But the "stroke" of the crankshaft is what drives ***port velocity.***

INDUCTION SYSTEMS

THE STROKE of the crankshaft determines ***piston velocity***. And piston velocity drives ***port velocity***. Why is this two-step fundamental worth such repetitive emphasis? It's a common mistake to equate a "stroked" crank gaining its extra torque at lower RPM because of "leverage" from the crank's longer arm. That's minimal. What's really happening is the piston is moving faster in the cylinder bore because the connecting rod that attaches it to the crank is following the crank around a much bigger circle in the same measure of time, which is by RPMs, and that's deceiving. The piston speed on a 4.00" stroke crank running at 6,000 RPM is the equivalent of 6,900 RPM for a 3.50" stroke crank showing the same 6,000 RPM. That's 15%. If we compare a 4.00" stroke 402 cubic inch engine at 6,000 RPM to a 3.50" stroke 402 cubic inch engine with .280" larger cylinder bores also running at 6,000 RPM, the stroked engine will be far more responsive at lower RPMs. It will also flatten out its peak power curve 800 or 900 RPM *sooner* than our big bore shorter stroke example. That's because the heads will have reached their maximum port flow velocity sooner. But there's another element: Not only is the piston speed increased in proportion to the increased crank's stroke circumference, but so too is its STOP and START velocity at Top Dead Center (TDC) and Bottom Dead Center (BDC) by a second factor called: **ROD RATIO**.

ROD RATIO is determined by dividing the connecting rod's center-to-center length by the stroke of the crankshaft. Most stroked engines will have *lower* rod ratios as a result of fitting the larger crank into a block that likely began life with a shorter stroke crank (like a Chevy 400 from a 302, which actually gets its genesis from the 265/283 CI design). Lower rod ratios come from reducing the connecting rod's length in proportion to the crank's stroke. The effect means it reaches a greater acute ANGLE to the cylinder bore centerline as it swings around the crank from perfect TDC or BDC, the only time the rod is aligned with the crank's stroke and there is zero angle with the cylinder centerline (C–L). Also, this shorter rod reaches its 90° relationship with the crank pin *later* in the stroke, which is *peak velocity.*

Detroit racing engineers determined long ago that the ideal rod ratio was 1.9:1; and "acceptable" rod ratios should never go below 1.75:1. Some rod ratios were over 2.0:1, for high RPM power curves. These precedents were set with the *BOSS 302 Mustangs* and *Z28 Camaros* in *Trans Am* racing when Ford and GM were both burning money to develop their engineering/racing legacies. Not long afterward (1978 or so) the Tractor/Truck pull crowd and IHRA "mountain motors" would begin to push rod ratios down into an insane 1.4:1 [and lower] range. This increased rod angle placed high side loads on the piston skirt and the cylinder wall, increasing friction and heat, which in turn distorted cylinder walls and required more piston skirt clearance. All this decreased the piston ring sealing efficiency to increase more blow-by. Some of these problems were remedied in-part by aftermarket blocks with increased *deck heights* to allow longer connecting rods, but the rod ratios were still lower than ideal.

With the piston's PEAK velocity being determined by the circumference of the crankshaft's stroke, *how quickly* the piston accelerates to this peak velocity is determined by the **rod ratio**. Lower rod ratios with their higher rod angles accelerate the piston to its peak velocity *quicker*, so the piston **stops** and **starts** faster at both ends of the stroke. The upside is a faster responding flow depression in the heads. The downside (besides the above) is higher inertia forces and stress, particularly on aluminum rod engines which see more rod stretch, and pistons and pins needing more meat. High rod angles also increase the severity of *reversion* waves in the induction system with some cam combinations, and this can tax the ultimate RPMs beyond whatever a higher rod ratio might have achieved. In short, higher rod angles (lower rod ratios) have a benefit for enhancing throttle response and accepting larger camshafts, *IF torque* is all you're looking for and where your application has no need of high RPMs. Like boats. But they have many side effects to efficiency and longevity if you step too far below their comfort zone: **1.75.**

INDUCTION SYSTEMS

Obviously, the effect higher piston velocities and lower rod ratios have on cylinder heads is significant. It makes the cylinder heads think they are running at a higher RPM than in *comparison* to a smaller engine with a shorter stroke crank. My two 400 CI engine examples above required different bores and strokes that would realistically be difficult (not impossible) to achieve in the same block, because the bores would have to be 4.280" to use a 3.50" stroke and have the same 402 CI as a 4.00" bore engine with a 4.00" stroke. But if my examples were traditional 350 CI and 400 CI small block Chevy engines where stroke was the only difference, you could argue the difference in "torque" was attributed to the difference in cubic inches. And to some degree that would be accurate, because you have more cubic inches working for you. But you'd likely see on a dyno the extra cubic inches from the increased stroke version would show more percentage gain of *horsepower*, but not as much percentage in torque.

The main point is that even with the two different engine sizes based only on stroke, the predominant influence of torque is coming from port velocity in the heads induced by the increased piston velocity of the crank. One perspective is a *symptom* and the other is the **cause**. Separating these are important.

Maintaining constant velocity of airflow through the intake ports is accomplished by what is referred to as consistent **cross-sectional area** ("**CSA**"). Wherever the CSA of a passage is reduced the velocity at that point will increase, and accordingly, airflow will slow down wherever the CSA is increased. The second element to velocity is **pressure**, which is reduced wherever the CSA is decreased, because the weight of faster moving air has less time to apply pressure; same principle of a carburetor venturi. In a mixture of fuel and air drastic changes in port velocity can result in "fuel separation." A passage that has been polished too smooth also leads to fuel separation. There is only one place on a maximum effort head where you want to reduce the CSA from valve size precedent dimensions: the ***short side radius*** and the ***valve bowl***. The short side radius is where the port floor turns for the valve (the *long side radius* is the roof of the intake port turning for the valve) is typically about 10–15% smaller in CSA than the upstream port CSA feeding it, which is a similar dimension for the *valve bowl* CSA.

It's easy when looking at an engine's ports to think they change shapes or are restricted in certain areas when in fact they may be very close to the same CSA throughout the head and manifold passage. This is illusive because factory engineers were forced to make the ports required size fit within certain parameters of compromise. Manifold height in some cases may be dictated by vehicle hood clearance, or pushrod location may be impaired by the "ideal" intake port location, and so forth. The port shape on most heads is the result these and many other considerations, including bolt locations, water passages, etc. The original small block Chevrolet head is a classic example. The intake ports appear to be tall and narrow at their intake gasket entrance, and then seem to be much smaller downstream as the floor actually rises somewhat to begin turning towards the valve. In actuality though the port is merely going from a vertical rectangle to a nearly square shape, and then to a slightly horizontal rectangle as it broadens itself to the width of the intake valve to aid in "pouring" airflow into the side widths of the valve. This produces a great, broad flow pattern over the short side radius that is consistent at both low and high valve lifts to whatever the peak flow flattens out at; and why it has been successful for so long.

I'll mention some of the fundamental rules for modifying any good competition head in a moment, but it is important to only make changes after you have measured the inside of the port accurately. Port molds are a good means, followed by simply wrapping a string around them every half-inch to get a circumference measurement. Naturally, a flow-bench will give you a great starting point, particularly to establish where the port drops dead at. See: *CHOOSING VALVE LIFT* (p.97).

INDUCTION SYSTEMS

BASICS: On small cubic inch engines designed to operate within a low to moderate RPM range the entire passage length of the head and manifold will be small, as will the valve size compared to larger engines; no surprise here. This type of an engine does not have as many high and low variables to the pressure waves within the ports as that of an engine designed strictly for racing; or even an engine designed for performance and normal duty. The latter is a good example of where a mix of manifold passage fundamentals is used to accomplish both goals with a minimum of trade-offs, as opposed to an engine designed exclusively for racing, or just chasing groceries. The families of engines we all know and love evolved from ho-hum applications; some from the 1950s; some from the 1960s. Then the factories took their fundamental platforms and updated them for hi-performance. The 396 and 427 Rat motors; the Z-28 302s and 350s; the various Ford "FE", 385 series (429–460 CI) engines, and the Cleveland series, to name only a few. Even the Chrysler Hemi began life as a groceries getter. The reason these engines are important to reflect on is the "standards" they helped establish that are still with us today, some fifty-plus years later. Even the most exotic engines born purely from the aftermarket with only vague similarities to these OEM workhorses, still owe their modern success to this history.

CONSTANT VELOCITY: This standard will never change, but how it is designed and used from mild to wild is specific. Let me edit down a long paragraph to a simple point, first: Think of a long funnel that has a 2" hole on its small end, and a 3" hole on its big end. Then think of a second funnel that has a 2" hole on BOTH ends. This second funnel is the physics of the engine described above, for stock use. The first funnel, is our professional race only naturally aspirated killer. There is NO constant velocity with this first example. This is a progressively accelerating velocity manifold and intake tract all the way to the valve. The second example is our *constant velocity* example. But here's the catch. This is theoretical "paper" values! In reality, our first example *IS* constant velocity **in effect**, because wet flow values of the air enhance *friction* and friction creates ***loss***. So we are compensating for this high velocity frictional loss by tapering the CSA. To be more specific, we are starting the manifold runner 10–15% larger than the valve CSA standard, then gradually tapering down to equal the valve CSA until the short side radius, then we are roughly 85–90% by the time we reach the valve bowl area. This is maximum effort, fast port velocity from big inches (stroked) or high RPMs; and of course the cam, compression, exhaust, etc.

On an engine that requires broader use power band, the runner formulas would be less aggressive because the velocity must be maintained at lower RPMs. A key point is the taper effect isn't need as much, because lower RPM broader torque curves are less influenced by the frictional loss.

There are general rules of thumb which are good starting points in relation to "how much" restriction or percentage of port size variations are allowed. For instance; the intake valve size is *ideally* determined by two things: (1) the cubic inches per cylinder which must be served, and (2) the RPM range which these cubic inches must operate within. The Detroit engineers have already established the fundamental valve CSA to port CSA with various cubic inch applications for a hi-performance engine, so I won't get too theoretical. But let's take an average run-of-the-mill American made small block or big block V8, with perhaps a tunnel-ram, or well modified 4bbl manifold for a strong 5,000–8,500 RPM. Let's also go to "air-flow 101" to begin with what the real precedent of any good system is: **THE VALVE.**

INTAKE VALVE: This is **THE standard** by which all ratios of port size and flow quality should be judged. The intake valve is *the last* doorway downstream of the INDUCTION SYSTEM. So how the valve's CSA compares to the induction tract's CSA is what we need to explore. Some rudimentary mathematical conversions are required which could be laid out more complexly, but I'll keep it simple.

INDUCTION SYSTEMS

FORMULAS: A modified small block Chevrolet might have an intake port measuring 2.22" high, and 1.00" wide respectively. To sum up the circumference each of these figures would be doubled (to account for the left, right, top and bottom walls), and then all four figures are added together for the sum of 6.44". This value is then divided by four (4) to arrive at equal wall lengths of: 1.61" "square." Multiply 1.61" times itself yields 2.59 sq. in. (2.59^2 in.). Of course now everything is easily on the web to pull up answers to any questions you have, like the next one; which is the area of a circle for calculating the valve area. But to remind you of your high school math, it is: AREA= Pi. X Radius Squared (πr^2). Our small block Chevrolet example would have **2.050" Dia. Valves**, and this works out to **3.302"**.

NOTE: I mentioned *port molds* earlier, which make all this easy. You can literally take a piece of nylon string (for small tight accuracy) and wrap it around the mold of the port at any location; mark it to get circumference; then lay it down to measure in linear terms, divide by four, multiply, and you've got your square inch value. You can also quickly see the bottle-neck(s) compared to the valve bowl diameter and the intake entrance. Same thing can be done with the Intake Manifold (with more foam).

In getting back to port/valve size ratios – after the valve size has been established in studying a system; the second point of priority is the area directly under the valve, commonly referred to as the *valve bowl*. This is the transition point of the short side radius to the bottom cuts of the valve job. Generally speaking the inside diameter of the valve bowl should be between ten and fifteen percent smaller than the valve. So, our 2.050" SB Chevy Intake Valve would have a valve bowl I.D. of **1.745"–1.845".** This will provide an ample amount of material for the bottom angles under the valve seat, as well as aiding in the airstream picking up velocity approaching the valve seat to enhance low lift flow. There is a critical hand *blending* of the valve bowl into the short side radius, where the corners of the port floor are brought to full width of the valve bowl, and the middle of the floor as it makes the turn actually protrudes inward (See: p.47). When done right, as you look into the valve bowl area with the head lying on a table with the intake parts down, there is a "D" shape (flat side down). And the corners of D would be our 4-o'clock and 8-o'clock positions of the valve bowl. This protruding radius is key to making the air "cling" to the port floor and follow it around to feed the valve area between 4 and 8 o'clock. When done wrong by excessive grinding the airflow actually ski jumps from the port floor to the valve stem, starving the near side of the valve. The only way to see this is with a pitot tube on the flow-bench. It has the effect of making your 2.05" diameter valve about 1.50". Anyway, this 10–15% smaller valve bowl provides this.

NOTE: Be careful **NOT** to confuse the 10–15% *valve bowl I.D.* to *Valve O.D.* formula as *AREA*. The "Diameter" Value is NOT the "Area" Value. For example: A Valve Bowl I.D. of 1.800" for our Valve O.D. is a *diameter* ratio of **87.8%**. But the *area* for 1.800" is **2.54"**, which makes the *area* ratio **77%** of **3.302".**

For a competition head, the head's intake port entrance should represent a very similar cross-sectional area as the valve head diameter. If a 2.05" intake valve is used on an engine such as the small block Chevy appreciates; then by using our formulae we can construct a circumference of 6.44". If the port was to be ground to a width of 1.00", that would require the vertical walls to be 2.22" high.

The next stage of priorities in working backwards from the valve diameter; is the intake manifold. Tunnel ram manifolds, and to a degree the open straight runner 4bbl manifolds have similarities which deviate from the rules of the more traditional "dual plane" intake manifolds. But generally speaking manifolds for racing have similar size ports as those of the cylinder head. Caution must be exercised: intake runners which approach the same CSA as the valve head diameter are usually the aforementioned tunnel-rams which are designed for use on high RPM engines with minimal low RPM requirements. Otherwise, dropping back to the 10% valve bowl rule would provide a broader response engine.

INDUCTION SYSTEMS

Engines that need the low RPM responsiveness that a circle track, or marine, or off-road competition engine usually requires, are better off with the intake manifold runner sizes using the 10% valve bowl diameter rule. As the RPMs may need to be higher, picking a percentage in between might see the best overall results. Understand that ALL this is part of the overall system, and I am not trying to give anyone a cut in stone SINGLE answer. These are solid general rules that you tweak depending on how serious your effort and *your budget* is. Dyno time works wonders for creating championship combinations. These formulas will get you in a very close starting position. in THESE applications, the valve bowl area is usually of the same size as the manifold runners, or smaller.

Before leaving this subject however, let me return to my funnel metaphor. The above talks about dual purpose, broad RPM and torque applications. For the maximum effort, full competition, high RPM (mostly) applications, or MONSTER competition engines, the Intake Runners will often require *more* port area than the Valve Dia. formula. A 2.400" BOSS Hemi Valve head can easily use a port that has the inside circumference of 2.400" to 2.500" (or more), particularly at the runner's entrance beneath the plenum. In such cases, the length of the runners (which I have not spoke about) are SHORTER.

I referenced "port molds" a couple times, and most pro engine builders are familiar with this. It is called "prosthetic foam" and is easily available on-line from a number of sources, but its origins of course comes from the medical industry suppliers. It is a two-part mix (a catalyst) that only takes four or five minutes to mix, pour and expand. Because of its elasticity it is easily manipulated from the port and can be sliced cross-ways at key points such as turns and bends with complicated shapes; then measured any number of ways from drawing a silhouette of the slices or string method I mentioned. ***IF*** you are fortunate to have access to, or can hire a facility to do a laser scan, then that can be converted to a solid model. This is the most professional route, and allows you to replicate the port with a 3D printer. This is nothing new; only the name and cost. For decades top shops did the same thing as "*stereolithography*."

Too much lift...

The intake valve and the camshaft which controls its lift is the "last outlet" of that heater duct we spoke of earlier. There is only a certain limit of airflow which can pass through the ports, just as there is with a carburetor. And this is irrelevant to the valve once you've opened the valve beyond the peak valve lift. If you were to take the valve out of the head completely, there would be a limit as to how much air would pass through the passages. What the valve does; is provide a "controlled restriction" to everything up-stream. This is sort of like keeping the slack out of a tow chain – in reverse. When the valve is being opened, it is like a small door which gets bigger and bigger. If you open the valve beyond what the port can deliver, then you've lost the responsiveness which signals the air in the passage further upstream.

One of the extra denominators separating my old heating system metaphor from an engine was the varying pressures, and the reversing direction of flow and force. I'll return to these one at a time. The varying pressures in the induction system are the result of the piston motion creating an instant depression or vacuum within the system. In terms of comparing the speed of the mechanical components to the "responsiveness" of the air itself, air is lazy. Just like skeet shooting, when timing the actual motion of the piston with the intended motion of the air column within the passages, a "lead" time is necessary. Hence opening the valves before the actual top dead center location is required, so that the lazy air volume "meets" the piston as it reaches TDC. On engines used for high RPMs these things are happening so fast that the air needs even more *lead* time. Since the piston accelerates from a standing still position, the rate which it initiates the depression is going to accelerate also. At these high RPMs this is occurring nearly 90 times per second to create high velocity cycles within the passages.

INDUCTION SYSTEMS

REVERSION in the induction system is the consequence of (1) air mass momentum reflecting off the valves when they stop the airflow column; and (2) piston reversal at BDC (enhanced by low *rod ratios*) when intake valve closing is too late. This second dynamic is when the piston works against the air column rushing behind the piston's down-stroke. As noted in **CAMS 104** (p.74) this is why the cam's *closing point* is the most critical of the four CAM events. When the cam's duration, lobe separation, and intake closing point is timed for the RPMs it is phased to work best at, the air column's volume and its inertia created by the piston will drive in so hard that it is essentially compressing itself beyond the static value of volume to surpass this 100% Volumetric Efficiency value. But *it is cam dependant*. When the piston stops that mass of air *can* bounce back *before* the valve traps it, and this is when the piston's efforts contribute to reversion. This reversion effect is initiated on the *piston side of the valve*. Its negative effect is mitigated by a *well chosen* cam. However, the valve's effect is another story.

For each piston cycle initiating the cylinder depression that air rushes in to fill there is a consequence when that air mass and velocity stops as the valve closes. That high velocity column of air mass in the runner tracts reacts to the valve closing like hitting a brick wall. Bouncing off of the backside of the intake valve, a pressure wave is created that reverberates back up the intake passage plowing head-on through the trailing air mass still in a whiplash momentum downward. This bad boy is called the **primary reversion wave**. Behind it a vacuum is created as the incoming air collides and fills in behind it. All this is happening in a few thousand microseconds that is repeated 65 to 90 times per second depending on the RPM. Constrained to the intake passage, it reacts to the *length* and *cross sectional area* variations to slow at larger points and speed up at smaller ones. When it finally reaches the **PLENUM CHAMBER** (p.139) or the ambient atmosphere, it expands violently and a vacuum wave rushes in behind it to provide a *boost* for the higher pressure (near ambient) ***secondary intake wave*** of air to rush in.

When the manifold runners are the right length for their area, and the RPMs and CAM timing is right for the head capabilities, then the primary reversion wave *will sync* with the secondary intake wave to aid the next cycle's airflow inertia and increase the volumetric efficiency. This "ram tuning" effect is optimum with naturally aspirated engines incorporating a plenum chamber. The primary reversion wave expands upon entry into the plenum area below the carburetors (or injectors) and works with the other cylinders to create a collective vacuum that is greater than when the RPMs are below the ideal tuning length. If the manifold entries are properly designed as they group closely, then each cylinder enhances the limits of the other's beyond their individual capability.

> **NOTE: To clarify a point:** *The "secondary intake wave" is really the boosted beginning of the main intake wave; not literally a second wave. But because the vacuum effect created by reversion doesn't sync until high RPMs – it is in fact a secondary phenomena that doesn't exist at low RPMs, and deserves distinction.*

This is really a simple observation, but the key with induction systems and choosing effective camshafts is to consider everything as a "system." The carburetor(s) (or injector(s)), the manifold, the cylinder heads – and whatever work that has been done – can still have a system "bottleneck." It may be the head's intake *or exhaust* ports; it may be the intake *or exhaust* headers; or it may be the carburetor(s) or injector(s). Matching the cross sectional area of the system and finding these bottlenecks is the secret. The cam is then chosen for a peak valve lift that is compatible with whatever limitation that bottleneck has in conjunction with the valve size being used, *which the flow-bench reveals*. It makes no sense to over-do one component if the corresponding system limits can't be harmonious with it.

See: ***CHOOSING A CAM* (p.99) and *CHOOSING VALVE LIFT* (p.97);**
See: ***LAMINAR FLOW* (p.133) and *PORT VOLUME* (p.140);**
See: ***RAM TUNING* (p.143); *SWIRL* (p.149); and *REVERSION, EXHAUST* (pp.143–144);**
See: ***TURBULENCE* (p.154) and *VOLUMETRIC EFFICENCY* (p.162).** *~jM*

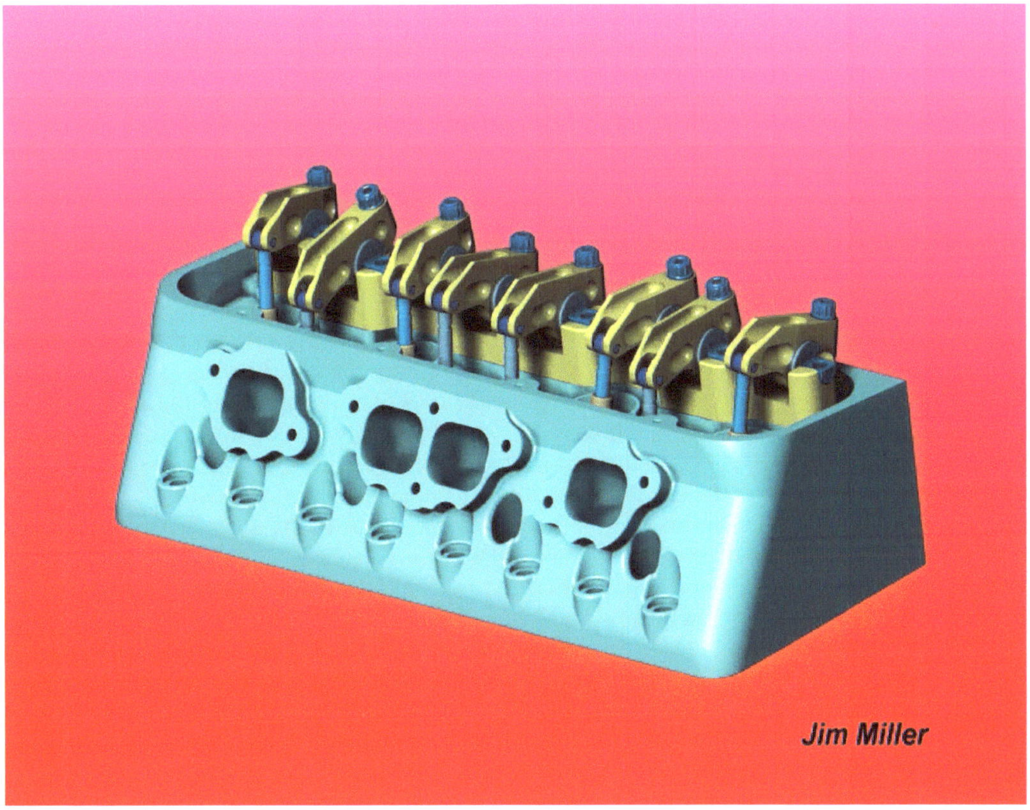

TERMS & DEFINITIONS

Racing & engine building have many terms and definitions. The following 52 pages of this chapter use minimal illustrations, but they do cover some of the most common and often confusing terms many of us have come to know, and a few that may be new for you. Including some of the confusing and overlapping terms that can make even the most experienced engine builder pause in thought for a moment.

The abbreviated version of this chapter originally released in 2003 was intended as only a summary of valve-trains terms. However, since valve-trains are so integral to cylinder heads, and problems in one can be so easy blamed on the other, this chapter was expanded for the 2010 Advanced Tech Bundle. The 2015 publication has been revised and expanded. Some of the terms that were previously "Tech-Tips" in this chapter have now been moved to their own chapters. These are: *Choosing Valve Lift; Choosing A Cam; Setting Valve Lash* and *Valve Spring Pressure*.

There are many rudimentary terms and definitions experienced engine builders will be familiar with, and perhaps even bored to find here; but I retained these in a "101" format as an courtesy to the aspiring who are just starting out. *We all started somewhere.*

Note: This chapter may contain some repetition of points from other chapters. For continuity most of these have been edited wherever possible, but some have been left intact for emphasis where relevant.

TERMS & DEFINITIONS

1/3 (ONE-THIRD) RULE: Refers to the traditional method engineers used for rocker arm design on an Over-Head Valve engine. See: **GEOMETRY, TRADITIONAL** (p.9).

ABDC: Acronym for *piston position* **AFTER BOTTOM DEAD CENTER**, measured in crank degrees.

ATDC: Acronym for *piston position* **AFTER BOTTOM DEAD CENTER**, measured in crank degrees.

ACCELERATION DYNAMICS, VALVE: The "net" valve lift, acceleration rate and duration (time open); or what the valve actually sees as a result of the camshaft's instructions – as translated by the rocker arm *design geometry* and its *installed geometry*. Since a cam profile is actually a blueprint for accelerating the valve-train from closed, to open, then trying to slow it down on the closing side, acceleration "dynamics" is changed by the rocker arm's geometry on either end. So when we add this term to the valve, we are really talking about the effect that rocker ratio and geometry have upon the velocity imposed at the valve, from what the cam's acceleration initiated. **See: VALVE LIFT DYNAMICS (VLD)**.

ACCELERATION RAMP (Camshaft)**:** Refers to the specific portion of a **CAM LOBE** that lies between the **BASE CIRCLE** and the **CAM'S NOSE**, which is precisely calculated to increase the **TAPPET** lift at a specific rate of acceleration, then slowing as the tappet rides across the cam's peak lift point, and begins to following the DECELERATION RAMP, where the compressed pressure, inertia and mass of the entire valve-train are hopefully constrained within operating limits of the desired RPMs, to allow the valve to seat with minimum bounce.

ADJUSTING NUT (ADJUSTER): Traditionally named on STUD MOUNT rocker arms for securing the rocker arm upon the rocker's mounting stud. Incorporates a SET SCREW which locks using an Allen wrench, to secure more tightly than OEM factory interference fit adjusting nuts when setting valve lash on performance camshafts.

ADJUSTING SCREW: Self descriptive, usually 3/8" or 7/16" **SAE** (fine) thread, with either a 3/8" diameter "ball" or a 5/16" diameter "cup" on the mating end for pushrod contact; used to facilitate valve lash adjustment on fixed height, stand mounted rocker arms (also known as shaft systems), or non-stud type mounting.

AFTERCOOLER (Turbocharger)**:** Often mislabeled as an "Intercooler," even by manufacturers and distributors of Turbochargers and accessories, the After-cooler is essentially a radiator, designed in one of two fashions, whereby it reduces the hot air temperature exiting the **COMPRESSOR** side of the **Turbocharger**, before being sent to the engine's induction system. Cylinder operating temperatures and fuel burn performance are more difficult with the warmer air, and they reduce the density of the air fuel mixture, and thus the engine's power. Coming in many shapes and sizes, the two basic design approaches of Aftercoolers is "air-to-air" and "liquid" cooled. In each case, the induction system air being forced into the engine from the Turbocharger's Compressor is routed through this aluminum "Heat Exchanger" that is comprised of a grid of small passages separated by, and attached to dissipating fins, just like a common radiator, whereby either (1) AIR directly passes across them for cooling, or (2) Liquid (usually water) flows across in carrying off the radiant heat. In the air cooled version, the Aftercooler must be designed and installed in such a way that the air passing through is forced and as cool as ambient conditions will allow, and of course separated from other nearby heat sources from the engine and/or Turbocharger. Sometimes a difficult task in tightly fitted engine bays. Water cooling versions, usually more complex, offer more latitude with such restrictions, but at the expense of more weight, so they are less common. Manifold tubing from the Compressor to and from the Aftercooler, should be designed with the same efficiency of airflow used on any induction system, with minimal lengths and bends being first goals. The efficiency of Aftercoolers pertains to how much reduction of the **Compressor Discharge Temperature (CDT)** can be attained, and in most cases the best of them will be in the 60% to 65% range. Meaning: whatever the CDT temperature is over the ambient **Compressor Inlet Temperature (CIT)** , approximately 60% of that *difference* can be reduced. This post-Aftercooler Temperature is known as the **Air Inlet Temperature (AIT)**, which is measured within the manifold runner between the Aftercooler and the throttle body of the engine, but usually as close to the throttle body inlet as possible.

> **NOTE:** Because CDT temperatures which often exceed 200° Fahrenheit make less power than cooler induction system temperatures that can use more fuel. These higher induction temperatures elevate the threshold for controlling detonation in the combustion chamber. Although the "boost" (and air flow) lost through the airflow's transition from one end to the other of a good, properly sized Aftercooler is usually about 2" of Mercury, most tests show the benefits from the cooler air alone outweigh this. Especially when many systems can be easily adjusted to compensate.

AIR INLET TEMPERATURE (AIT) (Turbochargers)**:** Sometimes confused as the temperature of air entering the **Turbocharger**, it is really the air entering the engine at the throttle-body. It is particularly the result of cooling the much warmer CDT (Compressor Discharge Temperature) air exiting the Aftercooler, on Turbocharged engines so equipped.

ANGLES ("101")**:** Is the fundamental division of a circle or any derivative thereof, whereby the measurement stems from a common axis or tangent point. A slice of pie's axis is the point where two edges come together. These measurements are referred to in DEGREES, MINUTES and SECONDS. Degrees is the full measure of a circle, and is divided into the equal numerical value of 360, while MINUTES and SECONDS are subsequent divisions having a value (as with a clock) of **60** for each degree measured. The symbols for this type of measurement is an apostrophe for MINUTES and quotation marks for SECONDS, and looks like this: 11° 20' 00". This is Chevy's 23° Small Block Head rocker STUD angle to the valve. Since the 20 minutes is even, the seconds value is "00."

> *A* ***second value system*** *for measuring the* ***fractional values of one degree****, (in lieu of minutes and seconds) is* ***DECIMAL****. In this second alternative to MINUTES and SECONDS, a single degree is divided by 10ths, 100ths, 1000ths and so on.* There is no division of the decimal measurement of one degree, merely the finer tolerances of more decimal locations. Our Chevy Stud angle measured like this would look like the following in decimal: 11.3333°. Depending on how many decimal of precision is used (10ths or 1,000ths), the more accurate terminology (and ORIGINAL) standard is still "Degrees, Minutes and Seconds," unless going to the fourth decimal place, but this is not usually done. When converting between one or the other remember that there is no higher value for MINUTES and SECONDS than "60." So a decimal value of "0.5" degrees is the equivalent of "30 minutes," or another way to see the comparative would be "45 minutes" equals "0.75" degrees. NOTE that the "0.5" and "0.75" both have the "0" in front of their decimal, to be sure there is no confusion that you are talking about a division of LESS than one degree. A small block Chevrolet's ROCKER STUD leans into the valve 11 degrees 20 minutes. The decimal equivalent of this is 11.333 degrees; because 20 minutes is 1/3 of 60, and 1/3 of decimal values (based on 10ths and 100ths) is going to be our "0.333" added to the 11 degrees. The traditional system of measuring angles in their full depth of Degrees, Minutes and Seconds, can get very specific. The ultimate division of one degree is found by first dividing by 60, then dividing each of the minutes by 60 again. This is the same as multiplying 60 X 60, which is 3,600. So you can divide one degree into 3,600 equal parts, by using the traditional

system. To get a finer measurement than this using decimal, you would have to take a decimal value out to the FOURTH decimal. Because 1/3600 is equal to "0.0002777". Which is a very tight tolerance when you're measuring angles. The DIGITAL PROTRACTORS we often use and sell for engine builders are limited to 1/10th Degree, compared to the more precise Protractors we use from Mahr, a German gauge manufacturer which cuts degrees into minutes and seconds. But the Digital Protractors of 1/10th degree will still be infinitely more accurate than anything any engine builder will ever require. Case in point: Our small block Chevrolet rocker arm will move about .023" for each degree of rotation. If you are using our Digital protractor atop your rocker arm to see what's happening in "angles" instead of linear measurements; then for each 1/10 indication of the Protractor, you've just witnessed .0023" (TWO and THREE TEN-THOUSANDTHS of an inch) motion. Is that accurate... or what? (**See: TOOLS)**

A totally unrelated point of interest: As the pilot of a vessel that flies or floats, you soon need to learn Nautical navigation, or at least some of its fundamentals. Among those who have, it is common knowledge that a Statute mile is 5,280 feet, and a Nautical mile is 6,072 feet. But one interesting thing that many navigators don't know about the history of this measurement, is where the actual dimension of a nautical mile comes from. They simply accept that a statute mile is 1.15 times greater than a nautical mile when they are converting for this. The earth is really cut up in measurements using degrees and angles much in the same division of measurement engineering uses. Even though it's not germane to terminologies of rocker arms and engines, there is a common element to this engineering method which has parallels. The very essence of a Nautical mile's length is based upon the equal division of the earth's diameter at the Equator. If you look at a nautical chart, or a simple world map (which often uses nautical references), you'll see our angular measurement shown in DEGREES, MINUTES and SECONDS. What nautical navigators learn is that the MINUTE value of navigation is equal to a Nautical mile. But what few realize until they think about it, is that the 6,072 feet was derived from slicing the earth's diameter at the equator from a 24 hour rotation into 21,600 equal lengths (360 degrees multiplied times 60 minutes); and this distance turned out to be our 6,072 feet, per slice (or our Nautical mile). The same value is given from North Pole to South Pole (Longitude), the Latitudinal lines in each direction from the Equator progressively show the minute values narrowing as they approach their apex. So the accuracy on a map for the minute "clicks" (as they are known) is only relevant at the EQUATOR and along the Longitudinal lines (North to South). You'll see both conventional and digital terms of angular measurement used with GLOBAL POSITIONING SYSTEMS (GPS), now found in our everyday aspects of travel. When dealing with something as big as the Earth, you can get more accuracy from a GPS by setting them to DECIMAL. If you set a GPS to the 5th decimal, you can actually cut the earth into roughly 6 foot spaces, which by coincidence, is roughly the civilian accuracy! All this isn't relevant, except at the core of many things a common standard exists, and sometimes it is interesting (if not beneficial) to understand what these common denominators are.

ANGLE FORMULA: A mean sine value of .0173"/per inch/per degree can be used to calculate rocker radial motion (when you don't have a computer handy). You can determine valve lift from degrees of motion; or degrees of motion from valve lift; or degrees of motion from cam lift. You multiply the pivot length (either the pushrod side or the valve side, depending on your goal) times a mean sine value of .0173"/per inch/per degree. The sum is then divided into either the cam lift or the valve lift, to determine the degrees of motion. If your quest is measured from the cam lift side, you must know the distance from the trunnion to the center of the pushrod cup. Multiply that figure times .0173" will give you how many thousandths of an inch motion will occur for EACH degree. If you need to know how many degrees of rotation the rocker will move, you merely need to divide your net cam lift by this figure. If your quest pertains to the valve side, then you need to know the distance from the trunnion to the roller pin centerline, and multiply that times .0173". If you already determined the rocker's rotational value in the pushrod side I just explained, then now you can determine what your net valve lift "should be," by multiplying the degrees of rotation times the sum of your rocker's valve side pivot length (roller tip center-to-center to trunnion). Don't be surprised if you find something not matching between "should" and "reality" with *everyone else's* rocker arms. Only a rocker designed to MID-LIFT geometry can make these values predictable and accurate. But remember, all rocker arms will have a "flex" compensation factor built in, if they are really accurate on net ratio, including a MID-LIFT rocker. The difference however, is MID-LIFT needs less correction than non-MID-LIFT designs, because they are correcting for geometry errors too. And thus, they lose precious area-under-the-curve. Cosine formulas are a constantly cumulative value based on each increase in angle, but you will find that .0173" inches, per degree, per inch of motion is a very accurate mean value. (Closer than you'll ever use on an engine.)

> EXAMPLE: Trunnion to **Push-rod Cup** Center-to-Center: .840"; CAM LIFT: .400". Multiply .0173" X .840" = .0145". To attain DEGREES of ROTATION: Divide .400" (CAM lift) by .0145= 27.58° (decimal); or 27°34'.

> EXAMPLE: Trunnion to **Roller Tip** Center-to-Center: 1.560"; VALVE LIFT: .750". Multiply .0173" X 1.560" = .0269". To attain DEGREES of ROTATION: Divide .750" (Valve lift) by .0269= 27.88° (decimal); or 27°52'.

ANGULAR ACCELERATION: Refers to the rate of change in the velocity of a device rotating around an axis, as measured from e specific perspective. In our sector (engines) this would pertain to the "perspective" of the valve centerline and the device would of course be the rocker arm. This term is sometimes associated with lengths of a device in comparison to another; such as a rocker arm that is longer over another. The term might be used to describe the degree of reduced angular motion of a longer rocker to impose the same net valve lift. The angle of motion for a longer instrument is less to impart the same amount of valve lift.

> **NOTE:** This observation is argued as beneficial to bearing reciprocation speed (since it is rotating less degrees), in and out arc at the rocker's end (because the rocker length/radius is greater), and the inertia values are influenced. These arguments for purposely extending a rocker arm's length are dismissed by one simple value, "aspect ratio." Because to make value of the longer rocker argument, you need to add more mass, which is more weight; and there is also more flex, because you are still asking a device with a similar HEIGHT (bearing bore and bearing bore housing area) to remain the same (to fit the head and stand), while you are extending its length to appease this failed argument of "longer is better." The ASPECT ratio of rocker body mass is now STRETCHED in the wrong direction (greater aspect ratio), and all these other detriments have now come into play: more weight, mass, inertia, flex. The **MOMENT of INERTIA** has now been extended further from its axial point (shaft centerline). None of which are a virtue to this argument. Next time you see the most advertised shaft rocker manufacturer arguing for his longer rocker theories, think about the above. *There is only **ONE TIME** when you want to stretch a rocker arm beyond what is necessary for hard part clearances to the rocker body (valve springs, retainers, etc.). We do this under a Patent Pending concept with the PRO-SHAFT™ and PRO-STAND™ systems.*

ANSI: Acronym for: ***A**merican **N**ational **S**tandards **I**nstitute;* a governing body of engineering standards.

AREA-UNDER-THE-CURVE: is a term which describes what a graph of the valve lift cycle would look like if charted on paper horizontally across TIME, in crank degrees of rotation, and vertically in thousands of an inch of LIFT. The quicker the valve opens, and "hangs" there, the greater the space beneath the curve from opening to closing. This "area" is what all cam and engine builders seek, because it maximizes cam efficiency. Unfortunately, the recording and studying of **VALVE LIFT DYNAMICS (VLD)** is rarely practiced by even the best of engine builders, until they learn how important this perspective is. Partly ignored because their previous tests are now inconclusive because there was no diligent record keeping of the **VLD**. Without practicing correct "**INSTALLED GEOMETRY**," it is never fully realized at the valve, Of course, you can't have this with a rocker arm that is NOT MID-LIFT in "**DESIGN GEOMETRY**" (p.12) either. You need BOTH: **MID-LIFT** rocker arms deliver this, but need to be installed correctly, which is why we've always included reference planes on the rocker arm which facilitate this for the engine builder. Technically, this term can apply to the cam lobe profile, too.

NOTE: Neither of the measurement references used for cams (lift and duration) represent proper rocker design or geometry when checking NET RATIO. Even the acceleration varies at the valve based on rocker geometry. Therefore, just because you measure another rocker system and find the NET valve lift to be close (or dead on) to the advertised value (using the cam lift to ratio formula), doesn't mean the maximum ratio with the least deviation is set throughout the rest of the valve lift cycle. That is because of the adjusting screw's position in the rocker. Traditionally, all other systems have had this too high, and when they didn't achieve accurate NET ratio, in later years (about 1995 and up), better known companies began moving the dimension in closer to the shaft (or trunnion) and placing a correction value into the ratio. The original problem of the adjusting screw (or pushrod cup) being in the wrong place and at the wrong angle was still there. So the rocker was slow coming off the early opening and trailed behind the possible velocity it could have had, until reaching its full valve lift. The exaggerated ratio, caused by this excessive correction factor on the mathematical "paper" dimensions for where the adjusting screw was at, gave everyone their accurate "net" ratio. But because the Design geometry was not accurate to the pushrod angles, the pushrod was leaving its linear path, wrapping around the rocker axis and thus being slow to lift the valve as quickly as it could.

Area-under-the-curve, can be seen, measured and compared by graphing out the acceleration rate of the VALVE on a piece of paper. Use graphing paper, and mark the vertical lines from left to right along the bottom of the graph paper to represent CRANK DEGREES of DURATION. Mark the horizontal lines from bottom to top along the left side of the paper to represent VALVE LIFT. There are TWO perspectives you can use, both achieve the same thing.

1. Select an even spacing of CRANK degrees, like every 5 or 10, and turn the crank over in these even amounts to place a dot where the NET valve lift is on your graph, or...

2. Select an even VALVE lift spacing, like every .010" or .020" and rotate the crank to open the valves, while noting the amount of crank degrees it takes to reach whatever your chosen spacing points is.

You'll get the same result either way, and the amount of gap between your chosen spacing is going to determine how specific and accurate your measurement is going to be; but it also takes more time with the finer jumps. Digital encoders and cam analysis software makes this easy.

Once you've "mapped" a particular setup, regardless of the rocker you're checking, then do the same thing again with a .100" different length pushrod. Doesn't matter which way you go, longer or shorter, just make the change and repeat the process. You'll see the effect AT THE VALVE (where it counts). You may also notice that there was very little change in NET valve lift, and ratio remained very close, in spite of this other loss at the lower valve lifts. What will you see? You'll see a significant difference in lost valve lift for the same amount of crank rotation; or put another way, you'll see it takes the crank more degrees of rotation for the same amount of valve lift. That's the lost area-under-the curve (valve lift "curve") that is being talked about, but rarely understood and even more often wrongly explained by both cam and rocker manufacturers.

ASPECT RATIO: This is a term used in a variety of forums, from engineering and physics to art. It simply means the divided sum of a dimension by an associated second dimension. I've used this term to describe the reality of support a valve guide's length gives to a valve stem in comparison to the off-center forces argued by some that are associated with the rocker arm not being centered in the middle of the valve. In this example, I cite the 2.500" LENGTH of a valve guide supporting a WIDTH of only .342", so the .342" is divided into the 2.500" for an "aspect ratio" value of 7.3. That's over SEVEN times greater support than the force applied to the width of this support. The main point made in this example is that it is not the off-center contact of a rocker's tip atop the valve that is critical (as long as it is entirely on the top of the valve), as much as it is the ANGLE of LOAD that is imposed by the rocker arm's arcing motion.

ASSEMBLY LENGTHS: Another big factor in determining your pushrod length involves any changes made to the head deck, valve seat depth, valve stem length, the block deck, and the cam's base circle. Different lifter heights must also be considered; for instance, you can NOT use the same pushrods for a roller cam, as that which was used for a mechanical or hydraulic cam – since the vast majority of roller lifters are much higher than typical lifters.

ASYMMETRICAL: The two sides of a concept or design which do NOT mirror each other; each having a unique profile, however still dependent upon each other for completing a task. The opposite of symmetrical. **See: SYMMETRICAL** (p.149). Asymmetrical and Symmetrical are two terms often used in CAMSHAFT terminology, referring to the simpler, older cam profiles which had even open and closing ramps, or the more current and most frequently used since the mid-1970s, asymmetrical profiles which lift the tappet quickly to open, and use slower deceleration ramps to set the tappet down in closing. **See: CAM RAMPS** (p.122).

ATTACK ANGLE: Refers to inclines toward or away from operating linear components, like the pushrod's incline towards or away from the valve centerline as it operates on an assembled engine. We also refer to this for the mounting stud or rocker pad's incline with the valve, which highly influences the range of valve lengths or rocker lengths to be used, as measured from the axis of rotation to the roller's centerline. In **CYLINDER HEADS** we use this term to define the **PORT WINDOWS** created with various port entrance and exit angles to and from the **COMBUSTION CHAMBER**. **See: ROCKER MOUNTING** & **VALVE TIP HEIGHT** (pp.47–48) for illustrations of the port roof and floor silhouettes relative to valves.

AXIS: The true center of any shaft or pin device in which a another device (usually bearing or roller) rotates about.

BDC: BOTTOM DEAD CENTER: Refers to the position of the PISTON (and crankshaft) in relation to the stroke of the crank. It is the ultimate bottom of the crankshaft's stroke before its travel back up.

BBDC: Acronym for *piston position* **AFTER BOTTOM DEAD CENTER**, measured in crank degrees.

BTDC: Acronym for *piston position* **AFTER BOTTOM DEAD CENTER**, measured in crank degrees.

BACK CUT (Cylinder Heads): Refers to an angle that is precision ground on the valve and behind the **VALVE FACE** angle, which is always a lesser value, and which is specifically intended to do two things: First, to adjust and reduce the width of the Valve Face angle to a specific width, usually equal (or close) to the **VALVE SEAT** width; and second, to blend the valve seat's angle to the valve head's even more shallow angle, so airflow across it at low (and high) lifts is less disturbed cross the Valve Seat.

BALL FUCLRUM: Refers to a simple "ball" principle of rocker arm mounting upon a Stud Mount principle to hold the rocker body to the cylinder head. The ball, which is actually a half sphere, operates directly upon a matching inside radius formed into the bottom of the rocker body to provide the rocker's central location to pivot on during its reciprocating operation. Such a design provides broad strength across more contact area for long life under OEM applications, where this design gets its roots many decades ago, providing it gets continuous flooding of cooling oil, usually from the push-rod. But its "friction bearing" principle was replaced long ago with various rocker arm designs graduating to some form of needle bearing. The irony to this design and its transition to needle bearing rocker arms, which <u>to this day</u> is completely overlooked, is that it has an inherent *self correcting, self centering* capability because it can "rock" on two axis, where the needle bearing rocker arm is constrained to only one axis of the rigid tolerances of the close fitting needle bearings. This dual axis freedom allows the Ball Fulcrum rocker body to *lean* to the side when needed, from the misuse (and incorrectly designed) offset rocker studs of In-Line Valve array cylinder heads, like the Small Block Chevy and Ford, which have increased port widths on most aftermarket designs, with corresponding offset push-rod locations. When the rocker stud gets moved off center from the valve to accommodate the offset pushrod, needle bearing rocker arms [unfortunately] get *twisted* (**see: Twisted Rockers**), and the rocker's roller tip does NOT sit squarely across the valve tip. The Ball Fulcrum rocker automatically senses this "compound angle" that has now been induced to the head design, and shifts its pivotal axis accordingly, even swiping diagonally across the new motion plane as needed throughout the valve lift. **MEI** recognized these virtues in designing its **OE Series** and **PF Series** chromoly steel rocker arms, and provided the necessary **slot width** to facilitate this side axial motion on such offset cylinder head installations, as well as provide using two different stud size ball fulcrums, without changing the rocker body.

BALANCE TUBE (Exhaust Systems): This term refers to the practice of connecting both sides of an engine's exhaust coming from each head through a common tube that crosses from one bank of exhaust to the other. Also used on engines having only two cylinders, such as motorcycles and other smaller powered vehicles. The main principle behind doing this is to capture negative pressure waves coming from one side to enhance a reduced pressure for succeeding waves that follow from the other side by more or less making them act as one. The fact that on multi-cylinder engines having two banks of cylinders, it is impossible (nearly) to design the system so that each cylinder is in the same sequence of timing ahead or behind another in how the exhaust is routed; often with two adjacent cylinders firing directly behind each other, instead of a more preferred and delayed spacing that will induce these negative pressures at a more preferential time. But the balance tube has proven to make the infrequencies of this cylinder timing less important, and in most cases enhanced improved power and flattened out the torque curve without hampering upper RPMs, as is usually the case with other tuning tricks. The diameter and position down from the head is important. (**See: COLLECTOR**)

BOLT STRETCH: Is the very specific and precise method of measuring the true value of a fastener's specific resistance in creating a locking force. It is a **TENSILE** measurement resultant of actually stretching a bolt to a predetermined length from its static (non-load) dimension, through the torque used to secure it. This phenomena occurs in all bolt tightening. Ironically, the tried and true torque wrench is actually measuring a very inaccurate (by comparison) means of stretching the bolt, because stretching the bolt is what the engineers and the principle really intends to accomplish. It is the only way to assure the proper pressure is applied, *where* it is to be applied, and to *stay* that way.

BRAKE SPECIFIC FUEL CONSUMPTION (BSFC): Refers to the efficiency of fuel consumption, per pound, per horsepower produced. The values need to include WEIGHT and TIME per Horsepower. It is acknowledged that 0.42 lbs//Hr. is a standard reference point per Horsepower. At 6.0 lbs per gallon of fuel, a given engine will burn 0.07 GALLONS of fuel per horsepower, per hour. At 400 Horsepower, that would be: 400 x 0.42= 168 lbs / 6 lbs (per gallon)= 28 Gallons/hr. Air density, humidity and temperature outside of standard conditions will vary this mean value.

BYPASS VALVE (Turbocharger): Sometimes referred to as a "Blow-Off Valve" (BOV), this is a check valve with a predetermined pressure limit designed to release excessive amounts of "boost" from the **Turbocharger's** Compressor feeding the engine, to prevent over-boost and engine damage.

ADVANCED ROCKER ARM TECH

Jim Miller

CAM: An eccentric shaped circle that has a base diameter which is less than 360 degrees that rotates about an axis which increases its diameter from the axis as it rotates through a 360 degree rotation. Generally egg shaped on designs that work with a direct contact running surface "cam follower," or having more squared radius` for designs where the cam follower incorporates a roller ended running surface. **See: CAM FOLLOWER** (p.120).

CAM BASE CIRCLE: (a.k.a. The "**HEEL**") The lowest point of measurement to the axis of rotation, this is the constant radius, NON-eccentric running surface of the camshaft that has NO effect on lift. It is also the mathematical foundation, as measured in degrees of rotation, for the starting and stopping points of where the lash ramps and acceleration ramps begin and end. However, these specific measuring points are gauged from a relative point, marked as a value of lift from the instant tangent point of where the ramps meet the true base circle. These points have been different over various cam companies and years of cam evolution, where tappet lift measurements such as .010", .012", .015" , 020" and so on were used to describe "gross duration." But the "standard" by which most real comparisons between cams is made, is marked at .050" tappet rise off "base circle." This higher tappet lift measuring point more equally compares two cams without the influence of lash ramps, which often can stretch out to look like a big cam, when measuring in this gross duration way, but in fact may be very tame by comparison to their .050" figures.

CAM CARD: The printed record of cam specifications provided with a camshaft to give fundamental design information necessary for selection, use and installation within an engine. This information should include **CAM LIFT**, **CAM DURATION** (often shown in two values, but most important is the .050" Tappet Lift), **CAM LOBE SEPARATION**, **CAM PHASING** (installed timing with crankshaft) and **VALVE LASH** (if mechanical). Additional information often included is "Theoretical" VALVE LIFT (based on specific rocker ratio, often not accurate to reality) and "Theoretical" DURATION (often taken at irrelevant lifter values close to the closed valve point, such as .012", .015" or .020" Tappet Lift; and usually of most value to hydraulic grinds, but often not considered of much value to experienced engine builders).

CAM CENTERLINE: Actually has two meanings. The usual meaning, refers to the cam's designed INTAKE centerline position, in CRANK degrees, as measured with the CRANKSHAFT position AFTER TOP DEAD CENTER (**ATDC**) of the piston. It is also known as the specification which cam "**PHASING**" is used. Again, the advancing or retarding of the camshaft with the crank's timing, as per the cam manufacturer's design specifications. (See: **CAM LOBE SEPARATION**) The second meaning, pertains to the more obvious implications of its name, meaning the actual centerline running through the base circle's axis, up through the cam's nose, thus dividing the opening and closing events equally. Even this though, can be misleading, since many **ASYMMETRICAL** cams don't have nose specifications that are evenly divided. None the less, there is a theoretical centerline to the computer's data which calculated and assembled the **CAM RAMP** dynamics.

CAM DURATION: This the period of "time" in which the **CAM LOBE** is active beneath the **CAM FOLLOWER**, as measured in one of two ways: degrees of rotation at the cam, or degrees of rotation at the crankshaft. The latter being the relevant standard for engine builders, and former being relevant to the cam engineer. Cam Duration historically has been measured from various, non-conforming reference points of tappet lift for its "*gross duration*," which technically is the total duration measured from the instant tangent point of cam ramp departure from, and returning back to the **BASE CIRCLE**. Gross duration has also been referred to on cam cards in different ways by various cam manufacturers who choose various starting (and ending) points for where this measurement is taken, always termed in thousandths of an inch rise from base circle; such as .012", .015", .020" or more of tappet lift on both opening and closing sides of the cam. The more this figure becomes, the less "gross" it is. So these higher measurements were more often used for timing purposes to install the cam with the crank by, or to compare one design to another for the effective duration (actually seen by the cylinder heads). But thankfully this comparison reference point has settled down in the last 20 years or so to including the more preferred .050" TAPPET LIFT reference. From .050" (fifty-thousandths of an inch) LIFT from BASE CIRCLE on BOTH Opening and Closing sides of the cam lobe, CRANK degree measurements are used to mark a particular cam's most active profile. To get some perspective on selecting cam duration for your engine. **See: CHOOSING A CAM** (p.99).

CAM FOLLOWER (TAPPET or LIFTER): A cylinder shaped component that rides upon the cam lobe of an **OHV** engine, and accepts a **PUSHROD** on the opposite end which provides the reaching link to the rocker arm atop the cylinder head. Cam followers are either hydraulic "roller," hydraulic "mechanical," or mechanical roller or straight mechanical, a.k.a. "SOLIDS." The ultimate performing cam follower is a SOLID ROLLER (no hydraulic valve lash preload); often simply referred to as a "roller." A roller lifter uses a hardened roller race which runs directly upon the cam lobe in conjunction with hardened needle rollers to create a true needle bearing design of high strength and impact resistance. It is permanently mounted within the tappet's housing with an interference fit solid pin, that may or may not have additional locking retaining rings set in grooves for added security. The ROLLER Tappet requires a *dedicated* geometry CAM lobe for both the hardened surfaces of the camshaft, but most particularly for corresponding with the angular acceleration formulas which the cam ramp requires to mate with the constantly rotating angle of attack the roller tappet has as it follows the cam to keep the acceleration dynamics of the tappet "in-line" progressive and smooth as pushes up from the cam centerline. The simpler way to explain this is that the angle of pressure by the cam lobe against the roller tappet is constantly shifting around the roller as it follows the cam lobe up and over the cam's nose, then back down. Whereby a SOLID tappet with its "fixed" surface lying directly atop the cam lobe is always receiving its upward lift in a straight line to the tappet's axis, so the cam lobe shape is much simpler.

> **NOTE:** *The velocity of the flat tappet lifter/cam is limited to a specific value by the DIAMETER of the tappet body and the BASE CIRCLE of the cam in conjunction with the LOBE LIFT being used. Since the rubbing surface on the TAPPET'S bottom side receives its "lift" by the nose of the cam sweeping across most of its entire face (contact surface) in the process of rotating beneath it, if the LOBE LIFT is too great or has too much angle coming off the BASE CIRCLE in its attempt to create more velocity, then the LOBE literally sweeps outside the diameter of the tappet. Accordingly, the tappet is no longer running atop the cam if this has happened, and destruction occurs.*
>
> *The ROLLER tappet does not have this problem because the point of contact between the cam lobe and the roller is always moving around the axis of the roller tappets roller centerline as it follows up and down with each rotation of the cam lobe. This concept allows nearly no limits beyond those established by the laws of inertia, mass (weight) acceleration and harmonics reaching their limits first.*

MUSHROOM TAPPETS (slang): Is a variation of the traditional, straight, one diameter dimension design that rides within the block's tappet bore, except the bottom "face" diameter has a greater "stepped" diameter that gave rise to the term "mushroom." The original mushroom tappets go back many decades where they were used on older, simpler engine designs with a smaller diameter main tappet body while still needing a broad diameter contact face to ride upon for the cam lobe. Racing applied the concept to the conventional design many years later as a loophole to get around "non-roller" tappet cam rules for racing which required only a flat tappet cam could be used, particularly in STOCK CAR (as it used to be called). The increased tappet diameter is a benefit only where the increased cam velocity and duration is needed for an aggressive flowing cylinder head, RPM and cubic inch combination, and only when ROLLER tappets are prohibited. Today, some classes of racing allow the mushroom tappet and some don't. The Chevrolet engines get around this rule limitation by increasing the Chevrolet .843" tappet bore to Ford's larger .875" diameter tappet to attain a higher cam rate. This mere 1/32" increase in tappet diameter can make all the difference for increasing the cam rate that separates a winner from loser on maximum effort record setting engine combinations. (See p.85)

CAM HEEL: See: CAM BASE CIRCLE (p.119).

CAM LIFT: This the actual measurement of the cam lobe's total rise from the **BASE CIRCLE** of the cam which is imposed on the **CAM FOLLOWER**. Cam lift on OHV engines is always a lower value of net **VALVE LIFT** because **ROCKER RATIO** multiplies this by 1.5:1 or more in most cases.

CAM LOBE: The eccentric portion extending from the cam's BASE CIRCLE to create CAM LIFT which comprises all three (3) stages of cam definition: LIFT, VELOCITY and DURATION, and also includes the transition ramp referred to as the LASH RAMP, which technically does not exist in of itself.

CAM LOBE SEPARATION: Pertains to the physical separation of cam lobe centerlines, in CAM degrees. This is also the reverse perspective of "OVERLAP." Both terms deal with the physical timing between the intake and exhaust lobes of a camshaft for the given cylinder in question. Traditionally, **OVERLAP** was the term used to deal with cam tuning to cylinder heads and cubic inch needs on a given cylinder head. In certain circles this still is. But LOBE SEPARATION became prevalent in the late 1970s as lower compression ratios became the norm from the Detroit manufacturers, adapting to more stringent EPA regulations. Increasing LOBE SEPARATION (which decreases overlap) is a way of building higher cylinder pressure on lower compression engines. See: Net Effective Cylinder Pressure (**NECP**). To get some perspective on cam lobe separation, **see: CHOOSING A CAM** (p.99).

CAM NOSE: The highest portion of the cam lobe eccentric where valve lift is at its greatest.

CAM OVERLAP: Refers to the time when both, the intake and exhaust valves of a given cylinder are open at the same time, measured in CAM DEGREES. Specifically, it is when the INTAKE valve begins to open, as the EXHAUST valve is just closing. The piston is still rising UP, chasing the exhaust valve closed, and the intake valve is opening to meet it. Increased overlap is needed on higher RPMs to provide additional time for the cylinder to "purge" itself of leftover gases from the prior combustion process. Physically, as explained in **LOBE SEPARATION**, the overlap cycle is the opposite perspective of the same consequence to cam design. They effect the engine's "running" **COMPRESSION RATIO**. You are merely saying the same thing from opposite perspectives. OVERLAP is a term not used much anymore, but was very much the perspective of cam development and comparison by professional engine builders during the pivotal cam development era of the late 1960s and 70s, where many of the breakthroughs of cam design were established. The transition to "lobe separation" lingo was the result of marketing by cam companies, beginning in the mid to late 1970s, who needed to boost cam sales and horsepower on LOW compression engines, resultant from the change in Detroit's gas guzzling car production, and ending the Muscle Car era as we knew it. **REDUCED** CAM OVERLAP is **INCREASED** LOBE SEPARATION. This reduced cam overlap increases **NECP** (**N**et **E**ffective **C**ylinder **P**ressure) within the cylinder, often boosting factory engine horsepower by in effect increasing the running compression on these milder, low compression engines. The greater OVERLAP was not needed on these low compression engines since they were not going to operate at high RPMs. Among advanced cam analysis in professional racing engines, cam overlap is still a critical perspective and term to be documented and used, particularly with cylinder head flow analysis.

> **NOTE:** *It is important to understand that you need to increase the static compression ratio of pistons (i.e., 9:1 up to 12:1) because a cam of greater overlap will lose "running compression" (NECP). Increasing the piston's compression ratio replaces this lost operating pressure from the increased cam overlap. (Assuming you have cylinder heads capable of breathing at these higher valve lifts and RPMs.) You cannot increase cam overlap and duration on a low compression engine without changing the pistons for higher static compression, and still achieve full potential. (Continued...)*

*Nor can you install larger compression pistons without changing to a cam of higher overlap (and duration), since now you will have too much "running compression" (NECP) and likely burn pistons, valves and create other problems from excessive heat, pre-ignition and detonation. The two run hand in hand. The only exception to these necessities being chosen only for higher RPM operation would be a large cubic inch engine with a large stroke crankshaft; as this increased crank stroke increases **PISTON VELOCITY** and increases airflow through the heads at lower RPMs.*

CAM RAMPS: Opening and Closing, as they are known, are the connecting portions of the cam's running surface that lies between the **BASE CIRCLE** and the **NOSE**. Sometimes referred to as the cam's FLANKS. The cam's **LASH RAMP** is also included here and designed to make the transition between the base circle and the more aggressive (quicker opening velocity) **ACCELERATION RAMP**.

CAMSHAFT: The multi-lobe shaft consisting of the required number of eccentric cam lobes needed to operate the valve-train, as it is timed by a chain or belt with the crankshaft. It is made of a variety of materials, with 8620 steel being the most common on high performance roller cam engines. Although numerous high tech coatings and other manufacturing methods are changing this norm on certain applications. NASCAR breaks the rules on cost and technology in this regard. **CAMSHAFTS** of today's high performance OHV (Over-Head-Valve) engine technology are the "heart" of a FIVE (5) component system that makes up the **VALVE-TRAIN**, which is arguably an antiquated principle of design, dating back nearly a century. Aside from the last component in our list: the "VALVE," these other three components are respectively known as the LIFTER (or tappet, or cam follower), the PUSH-ROD and the ***ROCKER ARM***.

CAM VELOCITY: Is an often used term by cam companies to describe, in a general way, the cam lobe's rate of lift over other profiles. It of course has a very specific engineering value to cam designers, which is given a value of thousandths of an inch *per degree* of either cam rotation, or crank rotation (double the cam value). "RATE of ACCELERATION" is another general term used for the same meaning, usually pertaining to cams; but we often use the term for *valve acceleration*.

CANTED VALVE: This term relates to engines which use a 3 plane valve centerline (not to be confused with valve FACE angles of valve jobs). The Chevrolet big block and 351 Cleveland and 460 Ford are just three examples which have their valves leaning on three separate planes, as compared to the cylinder bores of the engine block, for instance. Small block Chevrolets and 302 Fords, 340 Chryslers, etc., are on a "2 plane" valve array; meaning the valve's inclination to the cylinder bores can be described on 2 dimensions – with the 3rd dimension being zero – since they are all in-line with each other when viewed from an end perspective of the cylinder head.

CENTERLINE: Self descriptive, with rocker arms we are referring to the true center of trunnions, shafts, roller pins (and roller), bearing, valve and/or pushrod.

CENTERLINE, CAM: **See: CAM CENTERLINE** (p.119).

CFM: Usually cylinder head lingo; it's an acronym for CUBIC FEET per MINUTE. Most relative to AIR FLOW, measured through a passage, whether it is a cylinder head port, intake manifold, carburetor, fuel injection metering body, or whatever. It requires a specific pressure value for measuring, known as **FLOW DEPRESSION**. Without knowing the flow depression used to measure anything that has a CFM rating to it, such as cylinder heads, the value is meaningless for comparison. Too often, cylinder heads are promoted with a CFM rating acquired from flow testing, and no flow depression value is mentioned. So comparing to another head is useless and misleading.

CLOSING THE VALVE: Unfortunately, there are well known cam companies that promote an incorrect philosophy of rocker arm geometry; one of using the rocker for a tuning aid to the cam's acceleration rates. Cam acceleration is nowhere near its threshold of velocities which require the rocker arm geometry to be set in a way that the valve's velocity off the seat is artificially accelerated through what has creatively been called "accelerated ratios." Unfortunately, this is exactly what is promoted by some companies. Granted, there is no single way to do anything; and there may be tests which show an improvement in horsepower by such rocker arm juggling – but if so – it can only be due to a deficiency of what the cam design needs to be at a given lift value. Using the rocker arm to induce additional valve acceleration through juggled geometry makes properly controlling the test data impossible. In all cases for doing this, there is LOST efficiency and horsepower from the excessive motion the rocker and valve-train are now involved in. **See: OVER-ARCING** (p.137).

COIL BIND (Cylinder Heads): Refers to the collapsed dimension of VALVE SPRINGS. COIL BIND CLEARANCE is a necessary dimension that must be treated with exacting care and attention to cylinder head preparation. This minimum clearance varies with engine builders and some cam companies. Often taking the shallow side of safety. The type of valve springs chosen (single coil, double coil, triple coil) have specific cautions to what this minimum clearance is. Collectively, when all coils are assembled with the actual retainer to be used (because the steps for the inner springs can vary in height), this minimum dimension is acknowledged to be .050". *Trust me: do yourself a favor and make it .100".* **See: VALVE SPRINGS & PRESSURE** (p.90).

COLLECTOR (Exhaust Systems): This refers to the larger diameter tube which is used to "collect" the **PRIMARY TUBES** of an exhaust system's **HEADER**. Primary tubes usually range in size on V6 and V8 engines from 1.375" to 2.250", or more on radical and large displacement applications. While the "collector" will usually range from 3.00" to 4.50" (or more). The collector's main benefit is to give the high pressure waves exiting each primary tube a common chamber that captures and uses the low (negative) pressure following each wave to enhance the scavenging of succeeding cylinders before these exhaust pulses reach the higher pressure outside (ambient) atmosphere. (**See: VOLUMETRIC EFFICIENCY**) The collector is often where (or close to) a **BALANCE TUBE** is used.

COMBUSTION CHAMBER (Cylinder Heads): The cast or machined area of a cylinder head which (1) seals the top of the engine block's cylinder, (2) contains the valves (on OHV and OHC engines) and spark plug (or plugs) on non-diesel fuel engines, and (3) most importantly provides a specific shape that compliments the necessary blending of angles to and from the valves, as well as (4) controls efficient

FLAME PROPAGATION of the fuel ignition process. These shapes are varied, but follow two basic concepts dictated through the internal combustion engine's history that have been influenced by the type of valve and port angle configurations. These two basic shapes are known as WEDGE and HEMI (short for hemispherical). More advanced designs use more emphasis on cross flow dynamics that occur during the **OVERLAP** phase of valve timing, where exhaust port **FLOW DEPRESSIONS** are used with specifically shaped combustion chambers having various figure "8" appearances follow **SWIRL** concepts, rather than random or previously ignored flow turbulent flow of simpler designs.

Note: This chapter may contain some repetition of points from other chapters. For continuity most of these have been edited wherever possible, but some have been left intact for emphasis where relevant.

COMPOUND ANGLES: (Cylinder head lingo) Almost self describing, compound angles are the combination of three planes of dimensions (X, Y, and a straddled dimension between them ["Z" is the valve travel direction]. Most prevalent with *canted valve* cylinder heads, these include big block Chevrolets, 351 Cleveland and 460 Ford engines, among others. Compound angles are frequently used on pushrods, although the corresponding head may only be a 2 plane design; case in point: Chrysler Hemi. Here the valves are dimensioned in 2 planes, but the "offsets" needed for the pushrod clearance around the ports lean into a compound angle (compared to their respective lifter bores). **See: COMPOUND GEOMETRY** (p.54).

COMPRESSION: An external force applied to reduce the volume or dimension of matter or energy.

COMPRESSION RATIO: Is a static (fixed) measurement of cubic centimeter (cc's) volume (or Cubic Inch) displacement between the Bottom Dead Center value of a given cylinder's size (piston down), plus the volume of the combustion chamber and head gasket; in comparison to the compressed volume of the Top Dead Center value (piston up) of the combustion chamber (and head gasket) by themselves. This is ONLY a reference value for building and analyzing engine specifications with. It incorrectly presumes (to the novice) that valve timing to let air into the engine also opens and closes with the perfect top dead center and bottom dead center positions of the pistons. It of course does not. Because of delayed reactions from weight of air, mass, inertia, reflection and other dynamics, the valves open before top dead center and close after bottom dead center. So this theoretical perfect **TDC** and **BDC** formula is meaningless in heavy data acquisition of cylinder head efficiency and cam timing. The REAL compression ratio an engine runs at is more like 2/3 of whatever the STATIC definition here is. Known more accurately, as the Net Effective Cylinder Pressure (**NECP**).

COMPRESSION HEIGHT (Pistons): Used to describe the height of the piston's top deck from the wrist pin centerline. The piston's "deck" is the main flat TOP of piston that meets the piston's side where the rings are. On some pistons, such as hemispherical, there may usually be a protruding piston dome extending up from this tangent point, leaving no specific "flat" to be referenced. But the corner where this design reference resides is still the same, and compression height is a term all engine builders should know who work with piston manufacturers on custom designs, especially when calculating connecting rod lengths and crankshaft stroke combinations for engine blocks of a known **DECK HEIGHT**.

COMPRESSOR (Turbocharger): **See: TURBOCHARGER COOL SECTION** (pp.152–153). The cool section housing of a **Turbocharger** where the exhaust driven **Turbine** drives the compressor wheel (defined below) within the compressor, which then accelerates compressed ambient intake side air from an inlet side of the cool section through to the outlet side and into the engine's induction system.

COMPRESSOR INLET TEMPERATURE (CIT) (Turbochargers): This is the actual temperature as measured just before the **Turbocharger's** Cool Section inlet, coming usually from the air cleaner, which should be the same as ambient (normal outside temperature of the moment), unless influenced by a supplement system and/or heat radiated by adjacent components.

COMPRESSOR DISCHARGE TEMPERATURE (CDT) (Turbochargers): This is the actual temperature as measured directly after the Turbocharger's **Cool Section** outlet that is then fed to the engine's induction system, usually after passing through an *After-Cooler* designed to bring these temperatures back down by 50% to 60% of difference with the CIT. This increased temperature is the natural result of the Turbocharger's compressing of ambient air to a higher "boost" pressure.

CROSS-SECTIONAL-AREA (CSA): Head lingo, that refers to the actual cubic inch (usually) measurement of area for a port in a head, or intake manifold passage (or any other device flowing air, gasses or fluids). *Cubic Centimeters (CC's) is the other value of measurement, but rarely used in U.S. development, EXCEPT in the measurement of cylinder head combustion chambers, piston domes and cylinder's inch conversions to CC's with calculating NET static COMPRESSION RATIO.* The **CSA** calculation is NOT derived from a simple width times height formula. It is based from the square root of a value of area. A simple way to get it, which is good for aspiring head developers who slept through all the advanced math classes in school, depends on which way you are going to do the calculation. If you are taking a PORT and trying to find the CSA, you want to turn the shape into a CIRCUMFERENCE dimension; then DIVIDE by FOUR; then MULTIPLY this figure times itself. (i.e., A small block Chevy might have a port that is 1.25" wide, by 2.125" tall. There are TWO sides to EACH of these dimensions. So pretend you are a piece of string wrapping around the inside space of these dimensions and you will have a TOTAL LENGTH (circumference) of 6.75". Now, DIVIDE this by FOUR (4) to get our "square" for a square root multiple, and you have 1.6875". Now multiply this figure times itself (the essence of the square root), and you get: **2.847"**. That's the Cross-sectional-area. Now what if you are doing the opposite, and taking something that has a ROUND value already, and needing to find out the cross sectional area of it? Lets use the VALVE. Lets say our small block Chevy head has a valve diameter of 2.100". Here you need to use the standard, Pi (3.1416) to change this circumference into our "string" measurement wrapping around its circumference (not diameter). In this case that would be 2.100" x 3.1416= 6.59736. (To be exact.) Now divide this by four (4), and you get 1.64934. Multiply this times itself (1.64934 X 1.64934) and you get: **2.720"**. It's easy. All you are doing is converting shapes into a circumference, then dividing that into an even space of four equal parts, then you multiply one of those equal parts times itself. You are creating a four side box, and multiplying the width times the length. But this ONLY works for EQUAL sides of FOUR. That's why if you multiplied the 1.25" X 2.125" of our sample intake port, you'd have come up with 2.656", and this wouldn't have been the actual cross-sectional-area. It was 2.847", as shown above. In case you haven't noticed, I have also given you aspiring head developers a simple, easy way to check out ANY cylinder head, no matter how complicated the shapes. You need to use PORT MOLDS from prosthetic foam, available at hospital suppliers, and such. After you've made a mold of the port, you can wrap a string around various parts of the shape for your circumference measurement. Then do the math above, to compare with the valve sizes. Here's a rule of thumb. The port should be 90% to 100% of the valve's cross sectional area. With the **VALVE BOWL** area about 10% smaller on HEMI style engines, and 12%–15% smaller on Wedge engines. The intake manifold should follow a 95% to 100% value of the valve's cross-sectional-area too. This formula is for MAXIMUM EFFORT, high RPM engines. Torque builders would be considerably more moderate. All these variables can be their own book. My purpose here was to answer the question, then give those who really need to know this term a yardstick of how to apply it. Anyone not interested in cylinder head science isn't going to go past the first sentence of this definition anyway. **See: TURBULENCE** (p.154).

NOTE: Cross sectional area is the ultimate precedent for proper evaluation of port flow QUALITY. I've said it a million times, you can give a monkey a hand grinder and turn him loose making a sewer pipe out of a cylinder head port; and the flow bench will have (at some valve lift figure) a ridiculous **CFM** value. But it will MEAN NOTHING. Because QUALITY AIR FLOW as in "LAMINAR" (that always makes horsepower AND torque) comes from the greatest amount of airflow through the least amount of cross-sectional area port. This is where "VELOCITY" comes from, as well as controlling **LAMINAR** flow to prevent **TURBULENT** flow. Port shapes are not as important as cross-sectional-area; but they are important to controlling changes in cross-sectional-area. The only place you want turbulent air flow is in a carefully designed combustion chamber to control and confine it for optimum and fast flame propagation from the flash point. Turbulent ports aid this, which is why *wedge* chambers are still so popular. (But I digressed.)

CUBIC CENTIMETERS ("CCs", metric): A value we use in combustion chamber volume, piston dome volume, and deck volume during the process of determining static compression ratio. The CUBIC CENTIMETER to CUBIC INCH conversion formula is: **0.06102**; while the CUBIC INCH to CUBIC CENTIMETER formula is: **16.3871**. Unfortunately, CC's are also used to describe "port volume" which in the context they are used by many manufacturers and engine builders, is a useless value, as it is usually attached to the promotion or comparison of one head over another as if to correlate to "flow values," and this is incorrect and misleading. **See: PORT VOLUME** (p.140) for more perspective.

CUBIC INCHES ("CI"): Normally used to determine engine size. The formula for this is as follows: BORE x BORE x STROKE x .7854 x [number of cylinders]. After you calculate through to the .7854 value, you'll have the cubic inches of ONE cylinder; now just choose 2, 4, 6, 8, or however many cylinders the engine is to get the total value for the engine.
NOTE: The formal ANSI abbreviation for Cubic Inches is: "in^3" or "cu in".

CYLINDER HEAD: The essence of what an "internal combustion engine" is all about; the cylinder heads ARE THE ENGINE. This is of course one of the terms added to this section that created the opening note about including "rudimentary terms" to this section, and I hope if you're reading this, you've already got a good idea of what a *cylinder head* is! But I've added this definition mainly to emphasize that all decisions about valve-trains, induction systems and even cubic inches, as well as the end performance application for the engine... ALL come from the decisions made about the cylinder heads. Unfortunately, these integral components are often selected before the cylinder heads are chose or developed into their final modified form; with *valve lengths* too often being one of those choices. You CAN NOT decide Valve Lengths unless (a) you have a prior design to copy from and thus know the results of spring heights, retainer to rocker body clearance and effects of valve tip to rocker roller position; or (b) unless you have the head and you've checked all these things FIRST before ordering. DO NOT get locked into the sales hype of someone on the phone telling you that "100 long" is what you need. Don't buy any length valve that is longer than what you determine you really need; and you won't know this unless you've been careful about checking these other things just mentioned, FIRST!

DECK: Usually refers to either: the CYLINDER HEAD or the ENGINE BLOCK (cylinder block). It is the flat, gasket sealed mating surface which adjoins the HEAD and BLOCK together; and in many engine designs is a critical datum plane (measuring reference) for corresponding dimensions to either or both.

DECK HEIGHT (Engine Building): Refers to the actual height of the block's deck from the crankshaft centerline. But the term is also used to distinguish the "piston's" deck height beneath the block's deck when the piston is at **TDC**.

DEGREE-IN (a.k.a.: Cam degree-in, or degree-in the cam). **See: PHASING** (p.138).

DEGREE WHEEL: A circular plate made of any number of materials, being of anywhere from 4" and over in diameter, and precisely marked in equal divisions of 360 degrees along its circumference for providing exact measurement of cam timing when mounted upon the end of the crankshaft with a pointer that is synchronized to perfect TDC of the #1 Piston position. A necessary tool for all serious engine builders.

DETONATION (Cylinder Heads)**:** The more extreme symptom and next stage of PRE-IGNITION, and also the most destructive form of uncontrolled, nearly spontaneous ignition within the cylinder, which is not initiated by the spark plug. Main destruction is to pistons and rings, where the explosive detonation and cylinder pressures occur against the piston's rising motion. **See: FLAME PROPAGATION** (p.129).

DESIGN GEOMETRY: See: GEOMETRY, DESIGN (p.130).

DEGREES of ROTATION: A reference to rocker arms, whereby the rocker arm's reciprocating motion (typically referred to in linear terms of measurement) is referred to in terms of its angle of operating motion. Useful to calculating its critical MID-LIFT point from different perspectives of setting its Installed Geometry.

DUAL PATTERN (Camshafts)**:** Refers to different Intake and Exhaust Cam profiles. In the old days which stretched up into the early 1970s, camshafts commonly had identical Intake and Exhaust Lobe profiles ("single pattern"). Although custom grinds of dual patterns had been made long before the 1970s, it wasn't common. For the last thirty years, it is more uncommon to find a performance cam that isn't dual pattern than it is to find one that is.

DURATION: See: CAM DURATION (p.119).

DOHC (Head, Engine and Cam)**:** Acronym for **D**UAL **O**VER-**H**EAD **C**AM. **See also: SOHC** (p.147).

EXHAUST GAS TEMPERATURE (EGT): (Naturally Aspirated & Turbocharged) This is a critical temperature for analysis of cylinder air/fuel mixtures and tuning, per cylinder on any engine. Depending on the type of engine design, typical EGT temperatures with optimum air/fuel ratios and **BSFC** will hover around 1,450° Fahrenheit, give or take. Like any value, where the figures are derived are critical in comparing apples-to-apples; most data acquisition systems recommend that fast reaction probes designed for such temperature ranges be mounted uniformly at the same dimension from one cylinder to another, which should be not less than 4", or more than 6". Of course, unique applications may require innovation here. EGT is sometimes used in a less accurate way on less sophisticated engine instrument systems, where only a single probe is used in the final exhaust path, rather than on each cylinder, usually after the turbocharger outlet.

EXDUCER (Turbocharger)**: See: TURBOCHARGER TURBINE WHEEL** (p.153) for all related components.

RIGHT: The *EXDUCER* is the dual geometry blades illustrated on the compressor wheel to the right. These two blade designs are integral to the aluminum housing (not shown) in how the two stages of air flow work; one from the inlet (in the center of the housing), and the other in compressing the air and discharging it through the outlet.

ADVANCED ROCKER ARM TECH

Jim Miller

EXTRUSION (Metallurgy)**:** Technically, for our subject purposes an "extrusion" refers to the type of process from which the aluminum is formed. Extruding is the process of pushing hot aluminum through a die of a predesigned shape into bar lengths of usually twelve or eighteen feet.

This process is done for simple rectangular, square, and round stock, as well as nearly everything made of aluminum that has a linear dimension of a specific shape to it. Window frames are a good example of this. Because the shape is pushed through the die into lengths, it is important with some applications to consider the *grain* of the alloy runs linear with the length. With rocker arms, this process has been used for eons. So I am going to expand on this more than just the simple explanation you can dig up on Google. In applications where strength and longevity are a consideration, having the grain run in-line to the bearing loads, as it does with extruded rocker arms, means in extreme load and temperature applications fatigue and failure can follow sooner than if the grain ran crossways with the bearing loads. There are two other design options for manufacturing an aluminum rocker arm, which are argued to remedy this argument. One is FORGED and the second is BILLET. Forged is more limiting in manufacturing flexibility, and costly in up-front die costs, so even though a couple companies have went this route, it never really stood out as any benefit. Billet, something the author prefers for many reasons, is infinite in its ability to change and control all aspects of the final product. The argument against Billet by some "shops" is that for *production* too much material is thrown away. The author's response to this is that the material thrown away is (a) recycled and (b) insignificant to the other virtues for strength, flexibility and control. It also allows for the least amount of "grip" time to make the finished product, because the machine essentially cuts five sides of the part, all holes, all lightening and all finish cuts in one fixture mounting operation. Not to make a micro-article out of this subject, but extrusion, forged or billet are all subject to a good design of how much material is needed and where it is needed, as well as *what chemical composition* (grade) the alloy is, more than the technical arguments of which process is used. If the shape of the rocker arm does not comport to sound engineering for the placement of the three components that matter (pushrod cup or adjusting screw; the trunnion or shaft; and the roller), at the correct angles with each other, then extrusion, forged or billet is only bench talk.

FATIGUE: Is a term used to describe a real value to metallurgy and alloy properties, although it is not quantified as such. This is because it is the result of microscopic stress that are relative only to each component design, application and environmental operating characteristics. Specific designs of a product are subjected to cyclical tests, where the operating characteristics desired are exceeded in excessive endurance (time of use), operational stresses imposed, or both. Temperature is also a key component to the accuracy required to quantify the results. In *summary*: fatigue is the result of **stress** and **time** acting upon the molecular properties, where flex and temperature fatally disrupt the alloy's molecular integrity. It is especially susceptible and aggravating where machining flaws and **STRESS RISERS** can be found.

> **NOTE:** ALUMINUM is one of the best materials to use for a rocker arm, because of dampening properties; but it is also one of the few alloys (if not the only) which will FAIL from fatigue at operating loads which are

beneath its ultimate tensile and yield strength values. *This is the direct result of fatigue, which is not the case with other metals.* Other metals, such as steel, will operate indefinitely, if the operating loads are beneath their ultimate tensile and yield values, all else being equal. It is also worth noting here, that *fatigue in aluminum is directly influenced by the alignment of the* ***molecular grain,*** which is established during the shaping process at the mill; whether that is by forging or extruding. It is a generally accepted fact of metallurgy, that aluminum components which are designed to operate with their molecular grain in line with the load forces, will have up to SIX TIMES more resistance to fatigue. *It is also why all Miller Engineering PRO-STAND™, PRO-SHAFT™ and PRO-STUD™ rocker arms have ALWAYS been machined with their molecular grain running lengthwise to the loads.* While the biggest names in valve-train continue to use "cross grain" extrusion! (Sorry for the ad.)

FLAME PROPAGATION (Cylinder Heads)**:** Refers to the character of the burning process of compressed air and fuel in the combustion chamber following ignition by the spark plug (or plugs); but specifically: its progression from the point of ignition, across the piston and within the cylinder head's combustion chamber in a "controlled burn." Although the process is measured in milliseconds, it *is* a "controlled burn," and not an explosion, as some presume. When the burning process reaches a point of random ignition, where the incoming fuel is ignited by something other than the spark plug, either through other "hot spots" within the combustion chamber, or improper (and/or inadequate) fuel mixture, or some other unplanned form of **PRE-IGNITION**, the malady soon graduates to an explosion which is then called **DETONATION**. Once pre-ignition graduates to detonation it becomes destructive.

FLANK (Cams)**: See: CAM RAMPS** (p.122).

FLOW DEPRESSION (Head lingo) Used in relation to flow bench testing of cylinder heads, intake manifolds and passage ways. Flow depression is most often equated with "vacuum," and although this is usually the type of testing sought, it actually refers to either positive or negative pressure (vacuum) as measured in a relative of "differential pressure" from ambient (the pressure we stand and breath within and around us). In true scientific analysis, where everything must be standardized, tests done on machines which are not in controlled atmosphere (and most aren't – unless you're NASA), use a cross reference chart for all data derived in a test, which transposes the actual environment circumstances to what is known as STANDARD; a specific temperature and humidity value at sea level barometric pressure. Computer assisted data acquisition now does this in lieu of a chart, for the more sophisticated machines. Sorry, I digressed. But it is important to understand when comparing flow data between different tests, and especially different machines and/or operators, that without this cross reference standardizing, the results cannot be truly "apples to apples" in comparing. On high powered, high velocity machines, a MERCURY monometer is the reference of measurement; but in the real world of realistic flow test equipment, a WATER monometer (approximately 1/14th as heavy as Mercury) is used for witnessing the amount of flow differences being measured. This test "pressure" (positive or negative) is maintained for all points of valve lift, at a predetermined standard to operate the flow bench by in which flow data (**CFM**) is derived. One thing that should be noted here, is that FLOW DEPRESSION can be used in a way to exaggerate flow figures for anything being sold using these figures...like CYLINDER HEADS. All you have to do is TEST at a higher value, like 28" of Water Depression, over a test that is done at 25" and the flow figures will be higher. ALWAYS compare CFM values by the FLOW DEPRESSION, otherwise they are meaningless. It would be like measuring horsepower without knowing what the RPMs were for the test.

NOTE: The above directs the definition of FLOW DEPRESSION as used in "testing," but in engine operation, flow depression is more aptly limited to VACUUM, as that created by the piston and cam timing during normal operation, AND in the synthetic manner created by intake port and exhaust tube "tuning" (in conjunction with the cam and specific RPM ranges) where **REVERSION** waves, both super-sonic and sub-sonic create a flow depression behind them that can be synchronized to enhance depressions created in the intake system and across the piston for increased **VOLUMETRIC EFFICIENCY**. **See: RAM TUNING**.

FULCRUM: A one piece device used predominantly with the **O.E.M.** design rocker arms, consisting of a hardened, radius bottom ball or ball and stand combination (then termed a PEDESTAL FULCRUM). Having a hole through its center for a mounting stud that affixes the rocker arm to the head, the FULCRUM provides the rotational friction surface base in which the rocker arm pivots upon. Not desirable on high spring loads and high RPM operation, where its friction characteristics limit such use, as well as rocker arm design – since it is primarily a "stamped steel" rocker component. This design requires excessive amounts of oil flooding to adequately carry the heat of any performance use of such a design away from the fulcrum. Never use oil restrictors to the valve-train on performance engines using this design.

FULL COMPLIMENT: Bearing lingo, used in describing NEEDLE BEARING designs which have side-by-side needles (rollers). The other design reference opposite this, is NON-Full Compliment. This latter design incorporates a retaining spacer ring within the shell that acts like a cradle to keep the rollers separated from each other, with spacing that varies from design to design; but usually is about half as many rollers as FULL compliment. FULL COMPLIMENT is the most expensive, and is the higher standard of the two options, because of its inherent higher unit loading capabilities. Rocker arms having non-full compliment have made a compromising decision based on costs, usually not proportionate in real savings compared to the loss of operational integrity.

GAUGE HEIGHT/POINT: Is an engineering term for a specific point of reference to the design of something, often used for tangents to or from angles that need to be measured from a linear reference point. It is often used for many other things beyond engineering to reference procedures for inspection, manufacturing and assembly; especially where these other categories meet or overlap each other. We use reference to it here in describing the difference in measuring points needed for **VALVE LENGTH**, in discerning between measurements for designing your own length from the valve seat to the valve tip, versus the manufacturers usual term of overall length (**OAL**).

GEOMETRY, DESIGN: This is the geometrical parameters of which a specific rocker arm is designed, i.e., the dimensional distances between its connecting components – but specifically the "angles" in which they are placed within the rocker arm body; which are influenced by the operational angles of the pushrod to the valve on the engine in question. This correlation directly contributes to or from the translation of the camshaft's information to the valve. Because the tangent point locations and angle associated with the push-rod side of the rocker arm determine the degrees of motion the rocker arm rotates from whatever the cam lift is, we also refer to the Design Geometry as "push-rod geometry."

GEOMETRY, INSTALLED: This is the relationship of the rocker arm to the valve stem tip, as determined by the engine builder; or more specifically, the operating angle the rocker arm works as measured to the angle of valve. Although the rocker arm is in a "fixed" position during operation, which pivots about an axis that is stationary, the "perspective" of how it operates with the valve is determined by its pivoting relationship to the valve stem tip, or its shaft's height to the valve stem tip. Raising or lower this pivoting relationship directly causes the roller to move across the valve stem tip in varying amounts, known as the **WEAR PATTERN**, depending on where its "starting" point is at. If the rocker arms shaft sits too low to the valve stem tip, it spends more time rolling across the valve before it is pushing it "directly" down; therefore it is slower getting the valve off the seat. See comments for other consequences. **NOTE: See VALVE TIP ADJUSTMENT** (p.159).

GEOMETRY, TRADITIONAL: The **1/3 RULE** (See: p.9). Always based on two things: a SHOE TIP (scuff pad) contact surface for depressing upon the valve, and second: measuring the angles of motion from the contact surface of the shoe's contact pad. Measuring the pivot point's angular relationship with the valve through this contact point when the valve is approximately 1/3 of its total lift is how measuring the rocker arm's length and pivot position has been done since roughly when the Wright brothers went to Kitty Hawk. Incorrectly so, especially for roller tip rocker arms.

GROSS DURATION: See: CAM DURATION (p.119).

GUIDE PLATES: Referring to both valve-train and cylinder heads; these simple devices have been used for many decades to keep rocker arm alignment with the valve by restraining the push-rod's side-to-side motion. This concept of rocker arm alignment has often been simple "slots" cut (or cast) into a cylinder head, rather than the separate components that traditionally mounted beneath the rocker mounting studs, stamped from "plate" steel, and heat treated. They DO REQUIRE HEAT TREATED push-rods to operate without wear. *We highly recommend this concept of rocker arm alignment over the more recent, OEM cheap way out of using "self aligning" rocker arms. Avoid using self aligning rocker arms whenever possible, regardless who's brands of stud mounted rocker arms you use.* **SEE: Self Aligning** (p.146).

HARMONICS: Like music, harmonics denotes a self description of "frequency" and "vibration." Both are true in engines, because all matter has a natural harmonic frequency in which it resonates, all based from fundamental physics of mass and density, as well as material composition. For instance, aluminum has a different harmonic frequency than steel; and titanium has a different frequency than either of these. What is relative about harmonics in valve-train, is the interaction of these operating frequencies between the different components which establishes limits to RPMs, valve spring life and even necessary cam designs. Operational loads normal to a competitive engine compound once the threshold of harmonics is reached, requiring several times more effort to resist these increased loads. More simply stated, in valve springs, increased pressures for high RPMs is NOT added because of RPMs, it's added to replace the lost working pressure that harmonics destroys. Changing to a different frequency alloy, such as going from steel to titanium, elevates the frequency (because that is where titanium is at compared to steel: HIGHER) at which these loads compound, so less valve spring pressure is needed to control valve float with titanium valve springs. SIGNIFICANTLY LESS. Rocker arm geometry directly influences the threshold of any alloy harmonics, because their limits are a direct result of several factors, including "motion" and constrained linear loads held tightly to the linear paths of compression. In other words, *keeping the push-rods in-line as much as possible.* **See: *SPRING PRESSURES*** & ***HARMONICS*** (pp.90 and 91).

HEAD (Slang): **See: CYLINDER HEAD** (p.126).

HEADER: (Exhaust Systems) This refers to the main manifold system attaching directly to the cylinder head's exhaust ports for the purpose of routing exhaust gases from the engine. But "good" headers are designed so that the leading tubes leaving each cylinder, known as "**PRIMARY TUBES**," are of an equal length and specific diameter to accommodate a specific tuning of these high pressure, high temperature gases so that the energy created by their exiting the cylinder can be isolated from following cycles of the same cylinder, as well as used to enhance the scavenging of gases from adjacent cylinders, where each of these tubes come together in what is known as a "**COLLECTOR**." However, the term "header" is sometimes used with exhaust that have no collector, where only primary tubes exit from the cylinder head individually, as depicted in early designs on aircraft engines, boats, etc. Such designs still exist on supercharged engines, often used in drag racing, but are more often referred to in slang as "*zoomies*," because of their up-swept and short length that zooms backward with the flow of the wind. In *supercharged* applications the science of tuned lengths to cancel out **REVERSION** waves and match **NATURALLY ASPIRATED FLOW DEPRESSIONS** for high **VOLUMETRIC EFFICIENCY** doesn't exist, because the forced air induction of the supercharger pretty much overrides all these laws of ambient physics.

HEEL (Cams): **See: CAM BASE CIRCLE** (p.119).

HORSEPOWER: In its *purest* definition is: 1 HP= 33,000 Foot-Pounds of WORK Per Minute. In measurement, it is TORQUE multiplied by RPM divided by 5252. The essence of how horsepower is derived is based upon an understanding of the difference between WORK and TORQUE. FORCE is the common denominator of the two. WORK is DIRECT FORCE operating across a distance; while TORQUE is RADIAL FORCE operating around an axis. The formula of these is: POWER = FORCE x DISTANCE ÷ TIME. To calculate power, both DISTANCE and TIME must be derived, using FORCE across both of these. Where engines are involved, the rotational values of the crankshaft by its diameter is needed. To generate a formula for the TORQUE, the DISTANCE aspect is measured in RPM (revolutions per minute), times the RADIUS, times Pi (3.1416), times 2 (or 6.2832.). TORQUE is measured as Pounds of Force Times Distance. POWER is measured as FORCE times Distance per Minute. The 5252 value in our equation comes from 33,000 divided by 2x Pi (6.2832). Therefore: HP = Torque x RPM ÷ 5252. Something worth noting, is that at 5252 RPM, both TORQUE and HORSEPOWER will be equal. Beneath this value, Torque will be numerically greater, while above it Horsepower will be greater. Just buy a Dyno!

HORSEPOWER CLAIMS – Rocker Arms: Every rocker arm manufacturer loves to make claims, like "30 horsepower" from switching to their rocker arms. It doesn't matter which rocker or cam company you name, they've all made these claims at some point. We have NEVER made a horsepower claim when asked; even though it is very rare that significant power isn't made with changing to MID-LIFT rockers. But the ads run by different companies speak as if their products by themselves were responsible for whatever power was found, even if power was truly found. But it is not accurate, or we should say: it's not "apples to apples." THE PROBLEM with this, is that there has never been any neutralizing, comparative data to show that the tests were truly done in a way that isolates the substitution of their rocker arm from the previous example rocker, whatever that may be. In other words, there is never any data that accounts for the many other factors which influence what the engine sees in creating this horsepower, as a result of the rocker arm change. What difference does this make, you ask? Well, the ads present an improvement in performance, measured in horsepower, from a mere swap of rocker arms. But what is missing is that the change from a stock rocker to a roller tip rocker by itself, has a different lift dynamic at the valve, than the preceding rocker being compared to. In other words, the "valves" see a different valve lift RATE, probably duration, and maybe LIFT. In other words, the engine had a cam change, as far as the valve lift was concerned. This is where the credit for the horsepower should be placed. Not the rocker arm. So when you read these ads about "30 horsepower" improvement, remember that unless there is additional information stated that makes it clear a comparative test was done where the **VALVE LIFT DYNAMICS** was duplicated at the valve, then no credit can be fairly given to the rocker arm's features (roller tip, needle bearings, stainless steel, etc.), as the ads most probably imply. Remember, the rocker arm is DOWNSTREAM of the cam profiles, so whatever you do with rocker changes or geometry, directly affects what the valves see! That's what you're measuring. Now you might ask "so what?" You might say "I don't care how my rockers made more power, they still made more power!" But my response would be, that the ads are comparing THEIR rocker arms to OTHER rocker arms, and the OTHER rocker arms MIGHT have given you MORE power too...over the advertised rocker arms, if all else was equal. But it never is equal. That's the point. No one has ever used a "STANDARD" to cover all these variables. The point is, you have to be fair and careful of what you are buying, and why you are buying it. If you want to put more valve lift in your engine without a cam change, then changing rockers with higher ratios is a good way to do it, PROVIDING there is enough piston to valve clearance, and providing you don't lose running compression (from increased overlap) as a result, which then loses horsepower. You need to be clear in understanding the difference in TEST data that claims one thing, but which is the result of something else. That's all.

HOT SECTION (Turbocharger)**:** This is the heavy, cast iron side of the Turbocharger, which is mounted directly to or integral with the engine's exhaust system, often at the common collector for the engine's primary exhaust tubes. The Hot Section is carefully designed in its internal round shape and cross sectional area to feed the hot, high pressure, high temperature gases from the engine into the **TURBINE WHEEL** (p.153) at high speed, often from 40,000 to around 200,000 RPMs, which drives the **COMPRESSOR WHEEL** through a common shaft. It is not uncommon for the Turbocharger's Hot Section to get cherry red in many installations, and care to location and insulation (when possible) is needed in controlling this very high radiated heat. **See: TURBOCHARGER** (p.153).

INCLINED STUD: Most engines have a stud that is inclined to the valve's centerline. With the most popular engines, this dimension varies around the 10 or 12 degree range. That is, the rocker arm stud leans into the valve this amount. The engineers arrive at these figures through several considerations revolving around each engine's particular needs – but the over-riding consideration is to average (or split) the difference of the inclination of the pushrod into the valve. With the small block Chevrolet, the stud leans 11º 20' (11-1/3 degrees). The more you raise the valve stem tip, the higher the rocker arm must be; and the higher you raise the rocker arm – the closer the stud's centerline (and the rocker's pivot length) comes to the valve; making the juggling act of "centering" the roller to the valve nearly impossible from one engine to another. This is one major place (and theory) that most engine builders and manufacturers get into trouble. The logic of "centering" the roller on the valve is indicated as a good method for setting correct geometry, when obviously it's not.

INDUCER (Turbocharger): Pertains to the **COMPRESSOR WHEEL** (p.153) of the Turbocharger's Cool Section, but specifically it is the smaller diameter, and the first stage of two for the complex blade geometry of the impeller. This is the specific part of the compressor wheel which takes in outside air (usually from the air cleaner and connecting tube). After compressing with the corresponding inlet shape of the surrounding Turbocharger's cool side housing, it then forces it outward across larger diameter, integral blades of the same component which are known as the EXDUCER, whereby the air is forced into higher pressure and corresponding temperature (not a virtue) into the engine's induction system, usually after passing through an **AFTERCOOLER** (See: p.153).

INSTALLED GEOMETRY: See: GEOMETRY, INSTALLED (pp.16 and 130).

KEEPERS (Cylinder Heads)**: See: VALVE SPRING RETAINER** (p.158).

LAMINAR FLOW (Cylinder Heads)**:** Refers to the quality of airflow through a passage or over a surface. In cylinder heads, it is the quality of airflow through the ports, specifically the INTAKE port. Specifically, Laminar flow is LAYERS of smooth or spiraling airflow that is controlled and consistent. The opposite of this, is **TURBULENT** airflow, whereby air is reflecting and tumbling in all directions within the port passage. Turbulent airflow on a flow-bench can be heard loudly over laminar.

> **NOTE:** Experienced flow-bench operators who chart the airflow curve will notice one of two phenomena in the patterns of **low valve lift flow** up through the point where turbulent airflow overtakes this steadily rising increase until port flow drops like a rock. **The first** will be a steady rise in airflow percentage per step of valve lift chosen (.050" is recommended), followed by a gradually flattening out of this curve near the port's peak flow limit. When the valve lift area has exceeded the port's smallest area to feed it, the airflow will drop off quickly. **The second** example is a faster rising airflow, per valve lift, followed by a quick (without warning) drop, usually missing this leveling off of the flow curve near the top of its flow capabilities. This is the perfect example of a "turbulent" port; often too big for the valve it is feeding (or the valve is too small). When trying to apply the MEF (below), the POINT where a flow value should be taken is about 90% of the valve lift where flow increase STOPS. NOT turbulent, but where the curve is more or less flattened out. If that valve lift point is .750", and full turbulence is .810" (for example), then 90% of the .750" is what we've always chosen; which would be .675", as the CFM value used in the **MEF** calculations.

LASH: See: VALVE LASH (p.155).

LASH RAMP (Camshaft)**:** Is the area of the acceleration ramp where the cam lobe actually begins lifting the **TAPPET** from the **BASE CIRCLE** (or Heel) of the CAM. Although the lash ramp isn't really its own entity of the cam, merely the bridging area of the cam that has the slowest rate of acceleration to allow the increasing rate of tappet lift (in both height and speed) to continue to increase without too much shock that launches the tappet's contact away from the cam's surface. As noted below in *LOBBING the VALVE*, this was accidentally done in the late-1970s when a well known drag racer found more horsepower by excessive lash approaching .045". This was later pursued by other racing engineers with purposely radical acceleration ramps that were intended to "lob the valve." Both practices: excessive lash, and lobbing the valve with radical ramp speeds, are rarely done or accepted by professional engine builders, although a couple of "rogues" may still lay loose out there. But the most important aspect of both, was to change the limits of what controlled high velocity cam acceleration could be, from what was perceived prior to this turn of accident and events around 1979. The previous milestone of high performance cam breakthrough being when Chrysler contracted Crane around 1971 to develop a new cam for their Chrysler Hemi powered Pro Stock drag racing teams, and the .700" Valve Lift barrier had been broken.

LIFTER (Also "tappet."): **See: CAM FOLLOWER** (pp.85 and 120).

LINEAR MOTION: Meaning "in-line," linear is really self descriptive. Here, it pertains to the devices which are designed to follow an in-line motion (but often don't, because of improper rocker arm *design* and *installed* geometry). These two devices are of course the "valve" and the "pushrod," which work through the rocker's **RADIAL MOTION** (See: p.143).

LOBE SEPARATION: See: CAM LOBE SEPARATION (pp.70 and 121).

LOBBING the VALVE (Slang, Cam lingo): This is a technique of both cam design and application by the engine builder whereby a specific cam grind known to have acceleration ramps purposely designed with an overly aggressive velocity that will accelerate the cam follower and VALVE over the nose of the cam, literally throwing it into space at a given, terminal velocity from a predictable, high RPM, approaching the threshold of uncontrolled valve float. This can be said to be "controlled valve float." Its origins, like many evolutionary points in racing, was by accident, going back to Drag racing around 1980. It happened from Pro Stock racers finding horsepower by (accidentally) running increased valve lash; approaching .040" to .050" – a previously considered dimension that made valve-train destruction imminent. Which it is. But if used for very minimal periods of time (like drag racing), it can be tolerated – with risk. The data from this led cam "thinkers" to design ramps that would approach these previously taboo acceleration rates, but with more conventional valve lash. Rocker geometry is also pursued by some engine builders to induce these dynamics, but again, as stated elsewhere, there are tradeoffs for these second level methods of "tuning" which are unnecessary, considering roller tappet cam designs can still create whatever velocity is required (flat tappet cams have more constrictions – based upon tappet diameter), and the smoother, less damaging motions of the upper valve-train can be kept in check. Valve lash settings and lobbing the valve during specific controlled testing, should only be used as a barometer to indicate more specific cam grinds for the heads and performance application.

LONG SIDE RADIUS (Cylinder Heads): Refers to the top radius of the intake or exhaust port roof as it blends to (or from) the valve pocket in connecting the port's straight length passage way. These are often well *massaged* into a teardrop shape around the valve guide/boss to minimize its flow impediment.

LONG SLOT ROCKERS: Long slot rockers is a term used to define a modified stamped steel rocker arm body where the adjusting screw slot has been increased to accommodate a longer arcing motion of the rocker arm for more valve lift. The irony here, is why the manufacturers had to make them "long" from where they were, because they were long enough to accommodate the valve lifts being sought. But the real problem is that the slot was IN THE WRONG PLACE to begin with! It has often been more appropriately referred to as: "*Wrong Slot Rockers,*" for good reason.

MINUTES (Angular Terms), See: ANGLES (p.113).

MOMENT of INERTIA: Is the center of gravity equivalent to any mass which, in the case of rocker arms, refers to the starting and stopping effects of this mass in a reciprocating manner about an axis of rotation. The term is unfortunately misused in rocker arm promotion, as though its being "lower" is some accurate accomplishment of rocker body design. The "lower" moment of inertia which everyone so loosely throws around, actually refers to placing it closer to the center of axis of rotation. It is actually a "symptom" of design by our standards, because, unlike some educational entities, we don't design around it, we try to minimize it with the ultimate least amount of metal required to effectively control the unit loads and fatigue of a rocker body. No one who understood rocker arm design, would purposely add weight to a rocker body, except to correct some design flaw in strength. Moment of Inertia would "consequently" increase with this. But if that's what it took, then so be it. It is NOT a criteria for design, if all other aspect of body design follow minimum material for maximum strength needed.

MOTION LINE: Is the imaginary line to be made, drawing through the rocker arm's shaft or trunnion axis, and the axis of the roller pin. It is this line which is placed at a 90° angle to the valve centerline, at MID-LIFT.

MOTION TRIANGLE: Is a term we use based on the tangents located on the rocker arm, at the centerlines of either the pushrod side or the valve side which creates the THREE POINTS of critical geometry; TWO are from a connection between FULLY CLOSED to FULLY OPEN (cam or valve) and the THIRD is the AXIS POINT of rotation of the rocker arm. **See: MID-LIFT ARCING** (p.22).

MUSHROOM TAPPET, See: CAM FOLLOWER (p.120).

NATURALLY ASPIRATED (a.k.a. NORMALLY ASPIRATED): Meaning an engine which operates through conventional and traditional breathing created by the sequential cycles of vacuum induced by the pistons in timing with the valves, that allow ambient (natural) air pressure to rush in for compression and ignition. Non-assisted by supercharging, turbo-charging, Nitrous Oxide or any other external means of pressurizing of the cylinders.

NECP: NET EFFECTIVE CYLINDER PRESSURE: This is the REAL running compression ratio of an engine. It is the by-product of combined STATIC compression ratio, and cam/valve timing. As explained in **COMPRESSION RATIO**, the valve timing in reality, opens and closes beyond the perfect top and bottom stroke positions of the crankshaft, therefore not utilizing the full compression of the static compression value; usually about 2/3 or less (i.e., 12:1 CR = 8:1 NECP), but varies on the cam.

NET VALVE LIFT: See **VALVE LIFT** (p.156).

NOSE (Cam Lingo): See **CAM NOSE** (pp.60, 61 and 121).

OAL: "**O**ver-**A**ll **L**ength" refers (as it sounds) to measurements of linear objects, taken usually from their outside dimensions of one end to the other.

OE SERIES™: A deceptive appearing shoe tip rocker arm for competition engines needing a "stock type" (OEM style) design specific to limited classes of competition. Designed in 1997 for PVS, this design was unique as being the only stamped (forged) chromoly steel shoe tip rocker ever made (others are cast). But its real secret was a *simulated roller tip* for Mid-Lift geometry. A precisely ground flat atop the rocker provided the engine builder with a "*measuring face*" for setting the installed geometry, which was otherwise impossible since no easy axis points existed for measuring. Unfortunately the hardened steel limited tooling life, so only 40,000 units were made. Once released they sold out quickly at $128. Today, *used sets* can easily command *over* $400.

OEM: "**O**riginal **E**quipment **M**anufacturer" refers generally to the original automotive manufacturer, such as Chevrolet, Chrysler, Ford and so on.

OFFSET: This term is usually used to describe the difference of the pushrod alignment with the valve, as the head and rocker arm are looked at from an "over the fender" perspective (head's exhaust ports facing you). As intake ports are widened, there needs to be an offset of the pushrod away from the original port location with some head designs, to provide clearance. An offset guide-plate is the usual method with stud design rocker arms. To this day, it is still common practice to use stud mounted rocker arms which are "twisted" on the stud's axis. TABOO! (**See: TWISTED ROCKERS,** p.52) When this is done, especially on the small block Chevrolets and Fords which have 2 plane valve layouts, you are introducing a 3rd plane angle that will, at some point throughout the lift cycle, LIFT the roller off to one side of the valve tip's edge. (**See: COMPOUND GEOMETRY** p.54) The roller may lay flat in the closed valve position, but it will definitely leave this alignment during the valve lift cycle. You can see this on an engine that has operated as such, by a ring wear around one edge of the roller (indented ring in worse cases), and the rounding of the valve stem tip itself.

OIL RESTRICTORS: Is the technique of limiting the volume of oil to the top end of the engine through one of several methods, depending on the engine. One method restricts the oil feeding the tappet's oil galleys with plugs in the back of the block, another method simply uses restricted oil hole sizes in the pushrods, but does nothing for limiting a surplus of oil bleeding off from the tappets, if they don't need it. Some restrict in the rocker arm itself, with orifices down to .032" to .043" in diameter. The thinking behind this is to reduce a surplus of oil that is not needed to the upper valve-train, so that more is kept in the oil pan (or sump). A key method for doing this is in the main galleys which feed oil to the tappet bores. When Hydraulic tappets have been substituted for mechanical, the need for oil pressure is not as mandatory to keep the hydraulic valve within the tappet in a compressed state. This is reason number one. Reason number two, is the belief that excessive amounts of oil going to the top of an engine which is only used for short periods of time (such as drag racing), is a waste of resources from the crank's rod and main bearings. Reason number three, is "*I got needle bearing rocker's and I heard I don't need that extra oil up there.*" In most cases all three are wrong; and killed any warranty on my OE Series rockers.

ADVANCED ROCKER ARM TECH

Jim Miller

NOTE: The factory oil systems are fine on 99% of all the small block Chevrolet and Ford engines (big block too). The only shortcoming of the two is the gear-to-gear oil pump which Chevrolet uses. Ford's gyrator design is undisputed in self priming, high volume at low RPMs and stable flow at high RPMs. The history of "oil restrictors" dates back to Bill "Grumpy" Jenkins Pro Stock small block of the late 1960s and early '70s. Jenkins pioneered aluminum rods with loose clearances which slung a lot of oil off onto the cylinder walls and cam's underside. This additional bleed-off caused the top end to starve, eventually (in a few seconds) risking a void in supply to the high demand aluminum rods and main caps. By blocking off the oil through plugs at the back of the block which fed the lifter galley, Jenkins (and others) forced the surplus to be available for the crank. The ability of the oil system to pump oil through the lower-end exceeded the drain back capacity of the top-end, and since needle bearings in aluminum rockers needed little oil, the surplus was also unneeded. Somehow, probably because these "oil restrictor" kits got duplicated and hyped as a "must have," they crossed over into the copy-cat thinking of engine builders in all types of racing. If your engine is leaving too much oil in the top-end, revisit your drain back design. If you're using ball fulcrum rocker arms, which needed every edge they can get to pull heat out of that friction point, don't even consider this trick.

ALSO, I don't recommend you consider **TAPPETS** with OIL HOLES or GROOVES for *oiling* the **cam lobe**. As mentioned above, there is a full 360° oil spray coming off the side clearance of the connecting rods. This totally floods the bottom side of the camshaft, so the last thing you need to do is allow precious pressure and volume within the tappet that feeds the pushrod and rocker arm to bleed off on some waste of money gimmick like "cooling" lifters that add oil to the cam lobe. **These are a waste of money**. *~jM*

OVER-ARCING: Is a condition of excessive radial motion across the contact surface of a linear component driving, or being driven by a radial device – as measured by the perpendicular relationship of the radial device's axis to the linear component. In this case, we are talking about the travel of the rocker arm (either side) in relation to the valve or pushrod. Over and Under are opposite consequences, but the net result is the same. In over-arcing of the roller or rocker tip, it moves away from the rocker as it moves toward full lift; and on the pushrod side, the pushrod is moving out from the rocker axis toward the intake manifold. In each case the specific angle of rotation from the rocker arm is reduced, compared to effects of precisely dividing this arc with MID-LIFT geometry. In addition to this, the acceleration dynamics of the rocker arm upon the valve are influenced, with opposite consequences between the two. "Over-arcing," where the rocker arm is sitting too low in relation to the valve stem tip (the most common condition), will be slower to move the valve off the seat and begin to accelerate as it approaches full lift. "Under-arcing," where the rocker arm is sitting too high in relation to the valve stem tip, accelerates quicker upon initial opening, then slows down as it reaches full lift, rolling "inward" and UNDER itself. Both of these conditions contribute excessive side loads on the valve guide and valve stem, with subsequent additional heat, additional drag coefficients, increased harmonics in more exaggerated examples and reduced horsepower in all examples. These "symptoms" of bad geometry have often caused engine builders to increase valve guide clearances, change valve specifications and suppliers, as well as other alterations in pursuit of fixing nothing more than "symptoms," instead of the cause (bad geometry). Additionally, dedicated cam testing will be tainted by the lost information from this wasted motion, forcing engine developers to use more aggressive cam profiles to make up for the wasted rocker arm motion; adding more work to the valve-train with decreased efficiency and lost power. OVER-ARCING and UNDER-ARCING will impart LESS **RADIAL MOTION** to the valve (less valve lift), in addition to accurately mapping the cam's acceleration profile, compared to equally dividing these arcs with MID-LIFT geometry. **See: MID-LIFT ARCING** (p.22).

Note: This chapter may contain some repetition of points from other chapters. For continuity most of these have been edited wherever possible, but some have been left intact for emphasis where relevant.

OVER-HEAD CAM (OHC): Refers to an engine design whereby the camshaft lays above the cylinder head's valve array, and either opens the valves through direct contact upon them with the cam lobes, or through a leveraged action created by first pushing upon a single or dual end rocker arm, which then opens the valve with direct contact from one end of the rocker arm. In dual end rocker arms with OHC designs, the rocker arm may be above the camshaft, while on single ended rocker arms the cam would mount above the rocker arm. Either single or multiple camshafts may reside upon each cylinder head, controlling one or more bank of valves along the X-Axis (length) of the cylinder head.

OVER-HEAD VALVE (OHV): Refers to the engine design that has the valves over the cylinder block, even though it refers to a "head." This is an ancient and misleading acronym. Acronyms for other valve-train designs, like **OHC** (over-head cam), or **DOHC** (Dual-Over-Head Cam) are more accurate in their description. The main point that OHV means, is that the engine design has PUSHRODS and ROCKER ARMS. If you're wondering why Over the "head" is distinguished, it is because the prior design (which predates this author) was the FLATHEAD engines, that had their valves in the block. When the small block Chevy came out in 1955, the term **OHV** was born; and the Flathead was on borrowed time!

OVERLAP: See: CAM OVERLAP (pp.72 and 121).

PEAK FLOW (Heads): Refers to air flow through a cylinder head's intake or exhaust port. Specifically it is the point of valve lift where, as measured on a flow bench, the proportion of increase in air flow for the proportional increase of valve lift, as measured in "area," converge and go negative. RULE OF THUMB: When the increased value of air flow per 0.050" increase in valve lift reduces to less than 10% increase in CFM, the port is near death; and increasing valve lift beyond THIS POINT (NOT the "peak" where it loses air flow), is unnecessary forces on the valve-train, and thus lost horsepower.

PEDESTAL: This refers to the rocker arm mounting, usually of **OEM** design which incorporate a fulcrum that has a base to it which provides a solid mount directly to the head, whereas it uses a bolt instead of the older stud and nut design. The bolt is mounted directly through the pedestal/fulcrum and tightened until bottoming out. These are usually "fixed," non-adjustable valve-trains where all valve stem length dimensions, cylinder head deck heights, etc., must be set to factory specs. The hydraulic lifter is the only variable for minor assembly length variations of the valve-train; this variable is typically only .050".

PHASING: Is a term used in engine building which pertains to the synchronizing of the camshaft with the crankshaft; more specifically, the cam's INTAKE LOBE with the crankshaft's piston position within the cylinder, measured in crank degrees of rotation relative to the exact **TDC** (Top Dead Center) point of piston position. This measurement of the cam is often done in one of two ways, depending on the engine builder's preference. One, is by using the intake cam lobe's centerline for reference to the crank position; or two, the actual opening points on the cam ramps are checked with the cam card, dominantly chosen as the .050" lift position of the lifter. A dial indicator mounted and aligned precisely parallel with the tappet's motion is essential for measuring the cam motion, while a large degree wheel that has been accurately synchronized with the piston to indicate exact piston TDC is also needed. Advancing and retarding the cam from the cam manufacturer's published specifications is done after establishing accurate phasing first. A general analysis of a cam's acceleration rates can be done by using graph paper to plot the tappet rise and fall in 5° increments, but for precise analysis of the cam lobe, more sophisticated instruments with computer software are available to check the cam on a bench fixture or in the engine. But "phasing" is a term limited to simply timing the cam with the crank.

PISTON to VALVE CLEARANCE (PV): See: VALVE CLEARANCE (p.155).

PISTON VELOCITY: This is a relatively self descriptive term, but its real influence with engine performance is often not fully understood. Piston velocity, or "speed," is the result of two things: (1) Crankshaft stroke, and (2) RPMs (obviously). Piston speed is THE SOURCE for response characteristics of cylinder head air flow through the ports, in addition to the cubic inch *volume* each cylinder has with every stroke. These are two crucial elements to place in proper context: (a) Airflow VOLUME, and (b) airflow VELOCITY. Cylinder volume in conjunction with piston speed determines all of the tuning characteristics and specification needs within a given cylinder head's design, as well as the fundamental camshaft specs to meet these needs. These two elements can offset each other when choosing cylinder head packages and/or camshafts. In other words, if you have a large cubic inch small block engine, its increased stroke will achieve high torque at lower RPMs compared to a conventional size small block engine, but it can use larger port heads and greater camshaft specs that would more typically be chosen for a high winding engine of smaller cubic inches.

> **NOTE: Torque** is often the term used to explain what a large stroke engine achieves. Although it is true that the operating pressures created with each power stroke is applying leverage onto a longer "arm" for an engine having a greater stroke, and thus "torque" forces upon the crank are increased, the simple truth with what is happening in the cylinder heads with airflow is that **TORQUE** is a symptom from the increased PISTON VELOCITY that the stroke generated. The engine literally thinks it is going 8,000 RPM when the tach reads 6,800 RPM, because airflow and volume demand are greater at lower RPMs. There is a second dynamic to the piston's influence on the cylinder heads, dictated by a ratio of connecting rod length divided by crankshaft stroke. Known as the **ROD RATIO**, this determines how quickly the STARTING and STOPPING of PISTON SPEED occurs. But the ultimate piston speed is still determined by our two main factors (crank stroke and RPMs). The formula for determining piston velocity in FPM (feet per minute) is this: STROKE x 3.1416 x RPM ÷ 12 = FPM. If you use this formula when comparing different engine strokes, you can see how the RPMs are affected. TO COMPARE, pick your first stroke combination and use the foregoing formula. Then save the FPM value in your calculator and begin working backwards from this sum. Only now substitute the division symbol for a multiplication symbol, and divide where there is a multiplication symbol. SKIP over the "RPM" in this second step working backwards while changing the STROKE value to your second choice (i.e., 4.50" first calculation compared to 4.00" for second comparison), and your final answer will be the different RPM "effect" created in the ports by the different stroke. The answer to a comparison of a 4.50" stroke running at 6,000 RPM, compared to a 4.00" stroke is 6,750 RPM. In other words two identical engines of the SAME SIZE cubic inches having their only difference being stroke and bore, will have a 750 RPM increase being needed for the 4.00" stroke to have the same port velocity as the 4.50" stroke. Or more simply stated: the 4.50" stroke engine only needs to turn 6,000 RPM to make the same power that the 4.00" stroke engine can make at 6,750 RPM. When you consider that the stroked engine will also have more cubic inches (unless the cylinder bores were much smaller to keep the comparison the same size), the extra VOLUME dictates that PORT passages would also need to be increased, as well as VALVE LIFT. This is why it is easy to exceed the cylinder head capabilities when going into big stroke engines; not to mention all the increased work needed by larger induction and exhaust. When these limits are reached with such engines, the RPMs stop dead in their tracks, although you've created a "torque monster" at lower RPMs, which all starts with "piston velocity."

PIVOT HEIGHT: This is a term we refer to define the height of the roller pin axis above the trunnion/shaft centerline, when the valve is closed, which equals exactly half of the NET Valve Lift.

PIVOT POINT: The axis of the trunnion or shaft in defining the rocker body design. But also refers to the roller tip, when describing the height of the trunnion/shaft when pivoting the entire rocker body up or down from the roller tip, to establish **INSTALLED GEOMETRY**, as noted above in pivot height.

PLENUM CHAMBER (Induction Systems): A component of the induction system which is designed with or added to the runners of an intake manifold. It resides on the entrance side of the intake ports, adjoining the runners into this common open chamber of volume, which is usually beneath the fuel delivery, whether it is a carburetor or fuel injection system, although some designs of fuel injection systems with injectors having direct cylinder head mounting have been used with plenum chambers atop their runners. Usually however, individual runner designs with a single throttle body (butterfly) or independent throttle valves per runner are chosen. Its main purpose is two-fold: **(1)** To provide a "buffer" between the **REVERSION** waves coming back from the reflected flow hitting the closing intake valves, and avoid their disturbance of the fuel metering of the carburetor or injector's metering valve (on non-direct port injection designs). **(2)** Provide a larger "bank account" of air volume directly above the operating port runner that demands it; drawing from the collective supply of air flow through multiple butterflies that usually feed the plenum chamber, which otherwise do not offer adequate volume individually for the instant, millisecond demands that each cylinder would make at high RPMs.

> **NOTE:** When these reversion waves have not been arrested through a tuned length or a plenum chamber, they reach a frequency at high RPMs which actually escapes through the induction system's entrance; which in carburetors and some fuel injection systems creates a condition known as **"STAND OFF."** The phenomena can be seen as a fog or standing mist above the carburetor inlets and destroys power.

POINT-OF-INSTANT-CENTER: Refers to the tangent point where two intersecting lines meet, usually mentioned with angular devices which pivot and form linkage where this pivotal tangent point occurs. It is a phantom axis point. Pushrods that lean into the angle established by the valve centerline create a point of instant center several inches above the cylinder head which determines load force direction (induced by resistance to that force) upon the mounting locations of the rocker arm. A more appropriate use of the term applies to CHASSIS SUSPENSIONS, where 3-Link, 4-Link and 5-Link control arms, ladder bars and other similar devices mount in a way that run parallel to each other on one plane, with a positive attack angle on another plane that eventually creates a tangent point in space beyond the lengths of the component linkage itself, and having the effect as if they are this total length. A four-link suspension system on a drag car is adjusted so that if a line was projected out from both the upper and lower links, where they meet will have the same force of lift as if a more traditional ladder bar was used and its forward mounting point was at the same tangent point.

POLY LOCK: See: ADJUSTING NUT. (Nick name or trade name traditionally used in slang for **ADJUSTER**.) (p.112)

PORT VOLUME (Cylinder Head): This is usually referred to in cc's (cubic centimeters) of space. This is something that is nearly ridiculous to real cylinder head evaluation for port flow quality. It is a value, only relative to tuning or cylinder head and manifold design aimed at specific cubic inch engine sizes, where plenum volume in the intake manifold, RPMs being tuned for and engine sizes are all taken into consideration; and usually requiring computer analysis. Yet, this is often used to SELL cylinder heads, or EVALUATE cylinder heads, and this is NOT a good barometer for either. The only exception to this, is when the cc port volume of one given cylinder head combination is used against another by an engine builder who's quantified the "package" under a specific engine combination. This method of measuring a head is only good for calculated ram tuning 'theory." Often with computers. It has no sense in reality to really judging a good head from a bad head. For example, with a high speed hand grinder I can concentrate my material removal in one area of a port, thus increasing its "volume," while doing nothing to the valve bowl area, doing nothing to the intake port entrance, or do anything to the greatest bottleneck of the port (wherever that may be). But I have increased cc's, and the port is "bigger" by that measure. However, the FLOW BENCH (and Dyno) is going to show "turbulence" from the change in velocity as the air moves from a smaller cross sectional area of the port and into my new "ground out" area, where it will slow and tumble. Air flow "quality" is going to go down. Velocity is going to drop (from the turbulence), BUT... I am going to have a LARGER "cc" port. Another analogy to show how illogical this is, would be to compare a Ford head with a LONGER port, to a Chevrolet head with a SHORTER port, and of course the LONGER Ford head will hold more cc's of volume, so theoretically it is supposed to be "bigger?" I don't think so. But that's the goofiness that some companies and distributors try to sell heads by, or some gurus try to promote "tests" by. Forget "cc's!"

PORT VELOCITY (Cylinder Heads): Refers to the speed of air flow traveling through any given cylinder head's intake and exhaust ports; usually with more emphasis on the intake ports. This is due to the entirely different relationship between relatively cool intake port airflow versus extremely hot exhaust port airflow, which is really beyond using "air flow" as a term to describe violent high pressure gases escaping from the cylinder. With regard to the intake port, Port Velocity is the result of the ***Piston Velocity's*** influence on the port's cross sectional area. Even though its final value is determined by several other factors, such as cam timing, port length, reversion pressures and vacuums, on a naturally aspirated engine it is still ultimately limited to these two, fundamental components: ***CROSS SECTIONAL AREA*** (p.125) and ***PISTON VELOCITY*** (p.139).

PORT WINDOW: Used in Cylinder Head lingo, refers to the view one has down the intake or exhaust ports of a given cylinder head, where maximum view of the valve head is seen. **See: ROCKER MOUNTING** & **VALVE TIP HEIGHT** for additional illustrations that depict port silhouettes within the head, relative to valves.

PRE-IGNITION: (Cylinder Heads) A term used to describe the untimely ignition of fuel in the cylinder before the timed ignition by the spark plug; usually a result of white hot residual particles left over from the previous cycle, or an excessively hot valve, resulting from too lean or inadequate fuel mixture, and/or improper timing which increased operating temperatures.
See: FLAME PROPAGATION (p.129).

PRIVATE LABEL: A term used within the automotive aftermarket trade that refers to the sale of name brand products under a generic name, usually to a peer company or distributor who doesn't want the trade name used. Many companies buy products from name brand manufacturers, then repackage them, or have them packaged with their brand name on either the product and/or packaging.

PRO-STAND™: A moderate priced MEI billet aluminum stand mount rocker system (no longer made).

PUSH-ROD: The link between the **CAM FOLLOWER** and the **ROCKER ARM**; usually made of a single piece of steel tubing between 5/16" diameter on through to 7/16", they have many variations of length and end designs made to accommodate either a male or female radius end connection on both ends to provide freedom of misaligned movement while operating in a linear path. They have been made in different and exotic alloys, as well as shapes and end design combinations; but the net result and purpose is still the same: push the rocker arm up with the least amount of flexing.

PUSHROD CUP: The heat treated, pressed-in steel component of the aluminum rocker arm which mates with the ball-tip of the pushrod.

PUSHROD GEOMETRY: This is a term we've used to be specific in explaining that "rocker geometry" is based on the push-rod to valve angle. They are one in the same. **See: GEOMETRY, DESIGN**. (p.130).

PUSH-ROD HARMONICS: This is the vibratory motion that occurs at different frequencies and RPMs from the side-to-side motion, the rebound of compressed forces, influences from heat, friction and worse: high performance operational linear loads on a non-linear axis. In simpler terms: the pushrod misaligning with the tappet's centerline, and excessive in-and-out motion following a pushrod cup (or adjusting screw) around the rocker's axis more than it needs to. Like a worn tire on a car that shakes at certain speeds, but smoothes out at others, "harmonics" also occur in waves before reaching their terminal point. Inaccurate and over-arcing rocker arm geometry magnifies both their symptoms and effects, which we believe also dominos into early valve spring harmonics that then tend to feed each other. **See: HARMONICS** (p.131).

PUSH-ROD LENGTH: This dimension is critical. Accurate pushrod length is also very under estimated and misunderstood as to just how critical it really is. The fact that pushrods are sold by nearly everyone in .050" increments is testimony to how casually the valve-train industry treats this; when a fraction of this amount can influence ***AREA-UNDER-THE-CURVE*** and lost cam duration by several degrees. This is the essence of **INSTALLED GEOMETRY**; especially on the stud mounted rocker arms. Pushrods are the LAST components that should be bought for an engine, since their length cannot easily be established until the full valve-train has been mocked up with the cam, lifters and finished cylinder heads in the final machined engine block. On stud mounted engines the pushrod length provides the sitting height of the trunnion below the roller tip's axis when the valve is closed. This "sitting height" is measured in angles to the valve stem centerline by a line that runs between the trunnion axis and the roller pin's axis. Shortening the pushrod drops the tail of the rocker arm, increasing this angle and lengthening the pushrod raises the trunnion on the stud and decreases this angle. This angle is always going to be HALF of the total angular motion that the rocker arm will rotate for its given valve lift. (**See: FORMULA**) Very few engine builders will need to know, or bother with converting dimensions to know these angles, especially to set their installed geometry accurately, since we instruct and provide easier alternatives. This is simply done by measuring the height of the trunnion's centerline below the roller pin's centerline, and adjusting the rocker's height to be half of the net valve lift. (**See: INSTALLED GEOMETRY,** p.16-28) Because the dimension which influences this angle comes from the trunnion, but our adjustment comes from the pushrod, a required changed in trunnion height of .050" will require changing the pushrod by approximately whatever the rocker ratio is; or .075" for a 1.5:1 rocker arm.

> **NOTE:** If you haven't yet read the **History Of MID-LIFT™** and still wonder why pushrods NEED changing when factory spec rockers are changed to roller rocker arms, *even if the valve lift stays the same*, here is the summary: Roller tip rocker arms have "roller diameters" that vary from .480" to over .600". Factory rockers are "shoe" tip designs. The measuring point for what influences rocker geometry on the valve tip, is measured at the centerline of the roller, where in contrast, it is measured at the contact point on the bottom of the shoe tip design. Simply substituting *shoe tip* for *roller tip* ELEVATES this measuring point by HALF of the roller's diameter. So all else being equal, and assuming the factory rockers were anywhere close to correct for the valve lift, you would have to raise the rocker arm UP by HALF of the .480" diameter roller, or .240". But remember, this is AT THE TRUNNION. Which means you would need a PUSHROD LENGTH that was 1.5x LONGER than that, or .360" Longer than stock! If you've increased the valve lift by changing cams, or increasing rocker ratio (or both), then the MID-LIFT POINT of valve travel will be LOWER than stock valve lift (further from the valve tip, downward), so this increased trunnion height (rocker height) will be lower too, thus offsetting this .240" dimension and .360" longer pushrod. If valve lift was increased .100", then the MID-LIFT point would be LOWERED by .050", so the .240" trunnion height over stock would now be only .190" and the increased length of the pushrod would only be .285" (don't forget, it is changed 1.5x the trunnion height dimensions, so the value would adjust .075" less).
>
> Historically, because installed geometry had never been standardized before the principles of MID-LIFT were developed, its importance was ignored by the valve-train industry, so it is not uncommon to find that most rocker arm mounting studs are too short to provide an adequate number of threads above the rocker arm, once pushrod length has been increased for MID-LIFT geometry. The desired minimum threads above the trunnion should be 1.5X the diameter of stud. (i.e., 3/8" stud= 3/8" + HALF, or .562" height of threads above trunnion; 7/16" stud = 1.5x .437", or .655"). The very minimum should be no less than equal the stud's diameter. On STAND (shaft) MOUNTED rocker systems the pushrod length is just as critical, except rocker height adjustment is established by the stand itself; often machining its bottom to reduce rocker height, or "shimming" it UP to raise it. The adjusting screw on stand mounted rockers is kept to a specific length within the rocker body, usually determined by how many threads "out" from the cup being

fully bottomed out within the rocker. Jim Miller established this standard as "two threads out" back in 1978, and this has pretty much been adopted by all the well known "shaft" rocker system companies since. (*Current MEI design STAND mounted rocker systems use very specific overall length adjusting screws, flat washers and NUTS to control this fixed dimension even more precisely, which anticipates several different needs in setting geometry and tuning with lash adjustment, setting another precedent.*) Once the stand has been set to its necessary height, and the adjusting screws are set to two threads out, an adjustable pushrod is then used to zero out between the lifter and the rocker, PLUS valve lash is set, before a final overall measurement is taken to then make the exact length required. There is NO ".050" length averages used in buying pushrods here. At this level of investment and sought after performance, the pushrod length is ordered within .005" of exactly what is needed for MID-LIFT precision.

RADIAL ARC: As defined with rocker arms is technically a "symmetrically divided radius," which is the essence of the MID-LIFT principle. Because it is connected to a linear component and always measured perpendicular to the axis of rotation, its lateral motion is held to the ultimate minimum. (See: p.12)

RADIAL MOTION: Meaning "radius" (or circular motion), is the *general* measurement of any circular motion from a specific radius, but NOT relative to any specific lateral effect (sweep) or angle of incidence (attack angle). It's a generic term used in a general way to convey the difference from, and relative to **LINEAR MOTION** (See: p.134).

RAM TUNING: Induction system lingo. This refers to the design and tuning of an engine's induction system, and exhaust system, to achieve a flow depression across the piston that exceeds the standard vacuum it creates through natural operation. It is achieved with a concerted effort of exhaust manifold lengths, collector design and often cross-over tubes for the "initiating" aspect of creating this cylinder depression. But the cam and intake manifold are developed to compliment and complete the concept so a precise volume of air is waiting to fill the preceding and magnified flow depression in the cylinder that was created by timing the exhaust's supersonic reversion waves with: (1) the exact moment of peak piston depression and (2) cam initiation for the subsequent cycle. This is designed for a specific RPM for all this to occur. In simple terms: the vacuum waves created by the exhaust system are synchronized with the **FLOW DEPRESSION** created by both the piston and the **REVERSION** waves of the intake tract, so that inertia of the following volume of air (measured by port volume) can over-fill the cylinder. Sort of a naturally aspirated super-charging. **See: VOLUMETRIC EFFICIENCY** (p.162).

RATIO: Referring to *rocker arms*, is the difference in lengths between the rocker's axis of rotation and its two respective opposite ends, divided into each other. This is broken into TWO specific divisions of terms: **MECHANICAL RATIO** and **NET RATIO**. The mechanical ratio is the "paper" and actual measured ratio taken between these three points; while the NET ratio is the actual, multiplied "running" ratio imposed on the valve by the rocker arm AFTER FLEX, MISALIGNMENT, GEOMETRY MOTION and other real world factors have been included. All **MEI** engineered rocker arms are designed for **NET RATIO** at **ZERO LASH**, whereby on each engine we have already calculated a compensation factor into the MECHANICAL ratio to allow for these variables. Each engine is unique to these required compensation factors, and any deviations from MID-LIFT geometry will also change the NET ratio.
See: ROCKER RATIOS (p.93).

REVERSION (Cylinder Heads): Is a phenomena of reflected airflow which occurs in both the intake track and exhaust passages of an engine, predominantly in or near the cylinder head, but with the induction system, often within the intake manifold (or injection ports), and up through the carburetor or metering valve. It is the by-product of closing valves on the intake side, and subsonic and supersonic displacement of mass within the exhaust track. With the **INDUCTION** side, reversion waves are created within the intake ports by the reflected air that slams against the closing intake valve and bounces back up the intake passages. Proper runner length (and diameter) minimizes this by more evenly matching the VOLUME of air demanded by the cylinder; which is accomplished by the amount of volume contained in the runner and adjacent **PLENUM CHAMBER** if one is used. With the **EXHAUST** side, reversion is instigated and dynamic in a different way. More simply put, the super-sonic primary exhaust wave creates a vacuum behind it as it travels down the primary exhaust tubes. When it reaches the near ambient atmosphere it expands violently where a high velocity wave of air rushes in to fill the vacuum, traveling in "reverse" up the primary tube. This is the *primary reversion wave*. Limiting the reversion wave's affect entering the exhaust port of the cylinder head through various means, including internal baffles within an expanded tube of the header flange, have had some success on the Dyno. Port matching of the exhaust header tube with the cylinder head ended decades ago, opting for a deliberate step or "wall" of the cylinder head's exhaust port flange which inhibits the reversion wave's force.

> **NOTE:** The precise use of specific length exhaust tubes creates a timing of these waves within the cylinder which provide an instant low pressure atmosphere to aid in the succeeding cycle's exhaust process, but this is only tunable to specific RPMs.
>
> When done, optimum **VOLUMETRIC EFFICIENCY** can be achieved beyond 100%. It might be worth noting that we call this the "primary" reversion wave because there is a "secondary," and a third and a few more, depending on how bad the cam timing and port (and tube) dimensions are at a specific RPM of operation. These high **VE** results only come at specific frequencies when these subsequent waves reach what might be called a balanced frequency of the primary wave. At other points in the RPM curve, they can actually inhibit each other.

RETAINER (Cylinder Heads): **See: VALVE SPRING RETAINER** (p.158).

ROCKER ARM: Mounted atop the cylinder head in a variety of different ways, but always with one end atop the valve stem tip; the rocker arm is a "radial" instrument, designed to convey and transfer the "linear" movements of the pushrod and the valve, respectively. There are TWO (2) sides to the rocker arm, of which both sides "rock" back and forth about a common axis, usually pivoting on a "shaft," a "trunnion," or a "fulcrum." Yet, arguably, for more than 34 years of aftermarket engineering, it has been designed as though the valve is the only component to move, with well known companies teaching very bad and inaccurate philosophies to unsuspecting engine builders. One such fallacy, is to "use longer pushrods" to "place the roller in the middle of the valve stem" tip. Other trinkets have appeared over the years, supposedly to check this necessary pushrod length and set rocker geometry for you. But as you read a little further, you will notice there are many variables to setting a rocker in its correct location.

ROCKER DESIGNS: Every rocker manufacturer has their own reasons for any particular design of a rocker arm – but the truth of the matter is, most copy an already well used design; or worse, take what they think are good ideas from two or more designs and put them together. Without getting into mud throwing, which is not the intention of this information, let me simply state: there are so many variations between one company's rocker arms and another – for the same engine application, that using the logic of placing the roller on the center of the valve, will – more often than not – place the running geometry in a different state for each brand. I've even seen the SAME BRAND of rocker arm for the SAME ENGINE application have different lengths between the trunnion and the roller as a way to change the ratio.

> **NOTE:** On a more fundamental comparison about rocker arm design, something that should be pointed out when comparing a "shoe tip" rocker arm (OEM) to a "roller tip" rocker arm is: The trace geometry (path of tangent points) for a shoe tip rocker is entirely different as it goes through its arcing process to open the valve. With a shoe tip rocker arm, the contact point between the shoe and the valve tip is actually *stretching* and changing its effective length. How much change is determined by its installed geometry. You can actually mitigate (reduce) how much deviation there is, or increase it, simply by playing games with the rocker's pivot point height with the valve tip. But you're also changing how much and how fast the cam information gets to the valve at various points of the valve lift curve. With a **roller tip** rocker arm, this arcing motion is constrained to the "axis" of the roller tip, which rolls back and forth, but doesn't really

stretch its length as a shoe tip does. The comparison of these two dynamics is easy to see side-by-side, but is complicated to compare in a precision manner without dedicated cam and motion analysis tools and software. Let's just emphasize that the "rate" of change **at the valve** will be much different between a shoe tip rocker opening the valve, and that of a roller tip rocker on the same engine. This is also why I have always scoffed at cam or rocker companies horsepower claims being attributed to solely a change to their rocker arm. Even if a horsepower increase occurred, it was most likely the result of the valve motion seeing a bigger cam simply because of geometry change, and not from "roller bearings" or even an engineering virtue to the rocker brand over another, although such differences surely exist. Changing pushrod length to adjust rocker geometry when going from a shoe tip rocker to a roller tip rocker, all else being equal, is mainly this shift in measuring points that a roller rocker has over that of a shoe tip design.

ROCKER GEOMETRY: There are TWO kinds of rocker arm geometry: (1) DESIGN Geometry™ and (2) INSTALLED Geometry™. You can't do too much about "design geometry" because that is determined by the manufacturer of the rocker arm. But the "installed geometry" is the result of how you installed the rocker arm on the engine, which you do have control over. But be forewarned, all of the roller rocker arm manufacturers use different design geometry, because they never designed from any established standard; consequently, the adjusting screw or pushrod cup is in different angles and heights to the roller tip's location. When you adjust the installed geometry of the rocker arm on the engine, by using different length pushrods or stand heights, you will have different height pushrod locations that will move excessively in one direction or the other – depending on the manufacturer. The lesser of evils is still to have a 90 degree relationship on the valve side of the rocker arm, so that the valve and guide will suffer the least amount of side loading and friction. **See: GEOMETRY, DESIGN** and **GEOMETRY, INSTALLED** (p.130).

ROCKER HEIGHT: Is the dimension used to set accurate Installed Geometry, by measuring the rocker axis centerline (whether it is a shaft or trunnion), in relation to the roller axis (roller pin) in the CLOSED VALVE position. Whatever the NET VALVE LIFT will be as determined by the rocker's NET RATIO multiplied by CAM LIFT, the ROCKER HEIGHT should be HALF of this NET VALVE LIFT when the valve is CLOSED. As explained in the ***Installed Geometry*** page, all measurements must be in reference to a perpendicular angle with the valve centerline, and the valve spring retainer is the best reference plane for this. The first measurement, we've called **STACK HEIGHT**, accounts for roller tip above the retainer, while the second measurement finds the rocker shaft/trunnion's centerline in relation to this same plane, then added or subtracted as needed in meeting with whatever the NET Valve Lift dictates the MID-LIFT point to be.

NOTE: *This is especially easy when setting rocker geometry on engines with hydraulic tappets (roller or conventional), as it allows for accurate measurement of the required pushrod length WITHOUT collapsing the tappet's hydraulic valve from rotating the engine through its valve lift, as done when going directly to MID-LIFT. It is important to be sure the tappet's valve body (and pushrod seat) is "up" as it should be on tappets that have been pumped up from the engine's oil pressure, or from being primed. When using new tappets that have not been primed, you must be careful to set your adjustable pushrod gently to avoid any accidental preloading that would push the plunger down, and give a falsely long pushrod length. The tappet's hydraulic plunger motion is limited to about .050", but not knowing that it is fully extended when checking it this way is enough to lose several degrees of valuable cam timing at the valve.*

ROD RATIO (a.k.a. *ROD ANGLE*): Refers to the "connecting rods" between the piston and the crankshaft; but specifically, the mathematical division of a crankshaft's STROKE by the connecting rods LENGTH. Ranging anywhere from 1.3:1 (not good) to more than 2.0:1, the effects of rod ratio on an engine's performance and specifically its torque curve are significant. Rod ratio directly affects how the piston speed accelerates from its extreme positions of TDC and BDC (as well as slows down in approaching these points). The **PISTON VELOCITY** still reaches its same peak value, as dictated by the crankshaft's stroke; but this acceleration toward this is determined by rod ratio, with the smaller values of rod ratio making these quicker, and the higher values slower, respectively. One distinct down side of lower rod ratios, is their inherently higher thrust of the piston's skirt into the cylinder wall. These require more piston skirt clearance compared to a similar stroke engine using a longer connecting rod. Of course, point of fact: if you have two engines of the same model with the same stroke, but different length connecting rods, then of course the difference must be made up with either a different **COMPRESSION HEIGHT** piston, and/or a greater **DECK HEIGHT** block. It is commonly accepted that connecting rod ratios of 1.70:1 to 1.90:1 are preferred for most applications. Deviations above and below these values usually have some other negative offset. The higher the rod ratio, the higher the shift in RPMs for peak torque to occur. The more loosely used term of *rod angle* really has no literal meaning, since no one in engine development actually uses an angular value in describing or deciding what connecting rod lengths are used, not even in their effects of operation. See: *INDUCTION SYSTEMS* (p.104)

ROLLER: With roller tip rocker arms, we are referring to the roller which resides atop the valve stem tip. Mythically perceived to actually roll in operation, it does NOT. It merely provides a pivoting point for the direction of operational load imposed by the rocker body, much the same way as a roller lifter provides over a conventional "flat tappet" cam follower. With CAM lingo, we are referring to a ROLLER TAPPET. **See: CAM FOLLOWER** (pp.85 and 120).

ROLLER PIN: Sometimes referred to as the "roller axle," it is the device which the roller itself pivots around. Various manufacturing methods to attach this axle to the rocker body are used, from pressed-in (interference fit) to "rivet" or "nail head" designs which use a "C-clip" on the other end. At least one well known rocker manufacturer offers a needle bearing option. This has no benefit on MID-LIFT geometry. The well known manufacturer might see a benefit since their rockers have a high over-arcing friction.

SAE: Acronym for ***S**tandard **A**utomotive **E**ngineers.* An automotive standard of engineering used for many aspects of manufacturing dimensions and tolerances beyond the automotive sector. Fine threads follow SAE specifications.

SECONDS (Angular Terms): See: ANGLES (p.113).

SELF ALIGNING: Referring to rocker arms, is a rocker arm whereby the alignment of its tip with the valve is assured by protruding rails (on shoe tip designs) or washers (on roller tip designs) which straddle either side of the roller, and have an outside diameter greater than the roller for which they are aligning. The washer (or rails) purpose is to hang to the side of the valve tip and prevent the rocker arm from misaligning during operation. Standard roller tip rocker arms by numerous manufacturers jumped onto this OEM concept by simply juggling the roller width, or slot widths of their original "non-aligning" rocker arm designs to allow for these washers, which usually only extend anywhere from .030", to .050" or .060" along side the valve tip.

> **NOTE:** The so called "self aligning" rocker arm has ONLY ONE PURPOSE: TO REDUCE COSTS of MANUFACTURING and ASSEMBLY. *There is absolutely NO PERFORMANCE ADVANTAGE to this concept. It is deemed by MEI as a compromise which has risks to its use on any real high performance application. It is too easy for the roller tip to jump off alignment if any over revving, valve float, or other high RPM challenges come up during high performance operation. For this reason, MEI does not make the easy modification that we could in offering this option to our PVS rocker line. We strongly suggest that the addition of **GUIDE PLATES** be added to any application which had self aligning rockers from their OEM.*

SHAFT: Self descriptive, we refer to it here as the roller rocker arm's mounting type, rather than "trunnion," as used in "stud" mounted designs. Shaft mounting consists of one or more rocker arms mounted upon each shaft which is usually secured to an individual stand by two or more mounting bolts, or studs, on either end of the shaft.

SHAFT SYSTEM (Rocker lingo): Refers to **STAND MOUNTED** rocker arms, whereby the first person to coin this phrase was trying to make a distinction between rocker arms that use a removable shaft, from a rocker arm that has a fixed shaft, more aptly referred to as a "trunnion." In all reality, a trunnion (as pertaining to rockers) is a "shaft" and a shaft is a "trunnion." But this slang distinction stuck, and has been repeated for decades. We prefer "STAND SYSTEM" since it is a more accurate definition.

SHIMS: A variety of different strips of metal formed in one of several ways, and of many thicknesses, often custom made from sheet steel or aluminum. They've been around since the beginning of time for adjusting many things on an engine, during engine development; but with regard to rocker arms and valve-train, they have been used to change the height of rocker stands on shaft mount (what we call "stand mount") rocker systems. Because every rocker arm manufacturer we know of has always used stands that were too short for MID-LIFT geometry, they have always used this means to elevate their stands up to what each engine builder's requirements were.

SHOE TIP (Rocker slang): Refers to the fixed, radius contact pad of the rocker arm which lays upon the valve stem's tip, and directly pushes against it during the operational process. Primarily used on **OEM** applications for their simplicity.

SHORT SIDE RADIUS: Cylinder head lingo, that refers to the bottom side of the intake or exhaust port, where the bend to (or from) the valve pocket area is made to connect the port's "floor."

SHOT PEENING: A manufacturing secondary process whereby select, predetermined sizes of differing alloy "shot" is literally blasted by high pressure across selected surfaces of any ferrous or non-ferrous component, to microscopically *hammer down* and **work harden** the surface for increased resistance to surface flaws, fatigue and **STRESS RISERS**. Although there is no doubt that this process can enhance the surface integrity of some applications; its merits are arguably minor in most cases where adequate materials, manufacturing processes and engineering are properly applied.

SOLIDS (Slang, cam lingo): Refers to mechanical cam followers, also known as (a.k.a.) Solid Tappets or Solid Lifters. **See: CAM FOLLOWER** (pp.85 and 120).

SOHC (Head, Engine and Cam lingo): Acronym for **S**INGLE **O**VER-**H**EAD **C**AM. Refers to engines having a single camshaft above the cylinder head, as opposed to traditional engine designs whereby the cam resides in the engine block and uses linkage between it and the head (or heads) for opening the valves. **DOHC** is DUAL OVER-HEAD CAM (p.127), which is the same concept except for an engine design whereby the head (or heads) have two camshafts atop the cylinder head, usually dedicated to opening the intake and exhaust valve arrays separately.

SPLAYED VALVE: Cylinder head lingo, referring to valve arrangement with the cylinder head's combustion chamber, as measured in angles along the cylinder head's axis`. Specifically, "splayed" valve array is the leaning of the valves along the "X-Axis" of the cylinder (either end), making the valve stem centerline less than perpendicular with the cam's rotational axis. Rocker stud or pad mounting with this valve arrangement always requires a dedicated, compound angle to achieve accurate arcing trace in line with the valve's motion.

SPLIT OVERLAP (Camshafts): Refers to a point on the cam between the intake and exhaust lobes where the tappets (and valves) are lifted the same amount simultaneously. This is a confirmation reference point for timing with the Crank, once the integrity of each cam profile has been confirmed.

SPRINGS (Head slang): **See: SPRING PRESSURES** (pp.90 and 91).

STACK HEIGHT: A term we coined to define the height of the roller pin axis above the valve spring retainer. Used in setting ***Installed Geometry***, it is simply taken by measuring the height of valve tip above (hopefully) or below the top of the retainer, then adding HALF of the rocker's roller diameter in finding the axis. This measurement is then calculated with the second measurement taken from the valve spring retainer top to find the center of the rocker shaft (or trunnion) adding it or subtracting as needed. The final sum of the effort is known as the **ROCKER HEIGHT,** and this should equal HALF of the NET VALVE LIFT.

STAGGERED VALVE (Cylinder Heads): Refers to the valve centerline arrangement with the cylinder head's combustion chamber, where the intake and exhaust are shifted from each other along the Y-Axis (width) of the head. There is no specific criteria for this definition including or precluding the valve angles to head's deck, but it does mainly apply to valve stem angles which are perpendicular to the cam's centerline. However, compound geometry valve arrays, such as Big Block Chevrolet (where the valve's are "splayed" to the X-Axis) are often included in this term.

STAND: In rocker arms, it is the foundation of what the rocker "shaft" lays upon. In some OEM applications, it is cast into the head and it is more aptly referred to here as a Pedestal. Our description of other companies` "shaft" systems.

STAND OFF (Induction Systems): The condition of fuel and air that has been pushed back through the carburetor (or injector) inlets due to **REVERSION** and inadequate manifold design, to the point where a real and visible fog of this collective symptom lingers atop or at the opening of the induction tract or carburetors. **See: PLENUM CHAMBER** (pp.139–140).

STRESS RISER: A term used to describe the microscopic flaws that develop, usually in a corner, where a concentration of stress and flex will focus, and typically leads to a fracture. Stress risers are reduced or prevented by eliminating sharp corners with use of an increased radii where adjoining surfaces meet (rather than a sharp corner), and sometimes by micro-polishing. **SHOT PEANING** is another process that is used to "work harden" the surface of components that may suffer from surface fatigue, that can encourage stress risers to begin. Aluminum is especially susceptible to machining flaws that leave these corners where high loads occur, since aluminum is most volatile to **FATIGUE**.

STUD: This refers to any single bolt type fastening device, usually made of steel in varying grades, sometimes titanium, whereby both ends are threaded. However, in VALVE-TRAIN terminology it is the rocker arm mounting technique used with many engine designs, which we've always referred to as "**STUD MOUNT**." Specifically, the mounting of a rocker arm upon a "fulcrum" which resides on a stud mounted to the cylinder head, providing simple and quick production capabilities; although limited in strength and support for high valve spring loads and RPM use. Principally needed with "canted valve" cylinder head designs, such as the big block Chevrolets, 351 Cleveland and 460 Fords, etc.

> **NOTE:** Rocker arm studs are notoriously TOO SHORT for proper rocker arm geometry on many applications. This is a common fact, and too many engines are assembled with the set screws run down in the adjuster way too far because there is only four or five threads being used (and sometimes even less). In fact, with lesser valve lifts on milder cam engine applications, the problem is worse since the

rocker arm needs to be raised higher on the stud to accommodate placing the mid-lift point higher for the lesser valve lift. (Greater valve lifts move the mid-lift point lower in relation to the valve tip.) Do not be surprised to see that you only have a few threads left when you've elevated the rocker body up to set the installed geometry to mid-lift specs. You need at the very minimum an depth of threads that is equal to the diameter of the stud you are using, and ideally you will want 1.5X that diameter or more. Various companies offer "extended length" studs, but still some are not enough or they are too much.

One last point about ROCKER STUDS; as we all know they come in both 3/8" and 7/16" diameters. Many people switch to the 7/16" for more strength. But in MOST CASES when correct geometry has been used 3/8" is plenty adequate if using a strong alloy stud. The DOWN SIDE to 3/8" studs, is they are more likely to be TOO SHORT for correct MID-LIFT geometry.

STUD GIRDLE: Is a device made usually of two or more bars of aluminum having half moon cutouts in each bar to provide an elliptical grip upon extended length adjusting nuts when these two bars are tightened together by cross drilled and mounted bolts. Its purpose is to increase rigidity to rocker arm mounting studs on stud mount valve-trains, which would otherwise be free floating and flex under high spring pressures and RPMs. It dates back to the 1960s.

SWEEP: Used to describe the "roll" (or scuffing) across the valve stem tip.

SWIRL (Cylinder Heads): Refers to **COMBUSTION CHAMBER** design for cross flow that induces a preset flow pattern into the combustion chamber which is hopefully controlled into a rotating, high speed "swirl" that fills the chamber in a more complete but aggressive way, which does two things: first, follows the more violent patterns of turbulence created in a wedge combustion chamber, that has better **FLAME PROPAGATION** tendencies; and second: sweeps across the piston dome during overlap to scavenge the cylinder better. The piston dome, is viewed under these philosophies as the "floor" of the combustion chamber, and its shape too is often critical in the success of this often figure "8" shaped chamber. It made its debut across the aftermarket thinkers of cylinder head design in the mid-seventies, but is still a near black science to many. It does work, but its merits are really beneficial on high-tech, high-efficiency engineering, or aftermarket heads that have pursued this. Most cases of this latter example (aftermarket heads), the combustion chambers are compromised to follow some of these creative shapes, but still fall short of applying them to their fullest potential. Swirl technology depends on both: the intake and exhaust port *shapes,* as well as their ***ATTACK ANGLES*** *to and from a specific combustion chamber design, that works with the piston dome's profile.*

SYMMETRICAL (ASYMMETRICAL): With rocker arms (or cams) this refers to the equal division of motion, in mirrored fashion. Whatever the first half of the total motion is, is duplicated in reverse on the second half; more simply: "balanced," "proportional" or "even." "ASYMMETRICAL" is the opposite of symmetrical; having no balance or symmetry. With Asymmetrical, a term often used in cam designs, the opening cam lobe's acceleration profile is different than its closing side profile. This is done to get what is known as "area under the curve" during the greatest points of breathing capability, and then closing the valve where the most optimum point of ending this breathing is believed to be. Since the balance of these two goals is not symmetrical in proportion to the cylinder head's airflow characteristics, neither is most cams used in today's optimum performing engines, for either fuel efficiency or horsepower.

TANGENT: Is the term used to describe the actual meeting place of any intersecting lines of shape, form or physics. In rocker arm terms, we use this often to describe the axis points created within the pushrod tip or cup and the imaginary line that stretches between this point and the rocker arm's pivoting trunnion or shaft. Same for the imaginary line between roller's axis and the trunnion. Where these lines intersect is their "tangent" points.

TAPPET: Also "lifter." **See: CAM FOLLOWER** (pp.85 and 120).

TENSILE: Any force applied to or resultant from loads placed in an opposing direction (stretch). Opposite of **COMPRESSION**. It is also the most reliable form of predictable tension application in securing any component where maximum strength is required. In its ultimate use with fasteners, the actual stretch effect of **TORQUE** being applied is actually measured in its stretched value, rather than the torque of the applying force, which is less accurate.

THROTTLE BODY (Induction Systems): Refers to the controlling "butterfly" valve and housing which governs the air/fuel supplied to the engine's intake system tract. More specifically referring to Fuel Injection (FI) systems, where the timed fuel delivery is automatically controlled, in most cases by a computer and prescribed size injector orifices located in either the throttle body itself (not as efficient), the intake manifold or the cylinder head; which are synchronized with the degree of butterfly opening created within the throttle body that is linked to and controlled by the driver's "foot." In these latter examples of fuel delivery, the *throttle body* is merely controlling the air into the system. In all cases it is designed with an ultimate limited capacity based upon its inside diameter that equals the butterfly's diameter.

THRUST WASHERS (a.k.a. **THRUST BEARINGS**): Usually made of precision bearing steel, silicon bronze (or another suitable alloy as the application may require), these come in various dimensions from only a few thousandths of an inch, to whatever may be required. They are installed on the end of shafts or sides of components operating around a shaft to absorb lateral loads; often found on the end of camshafts, crankshafts, or other bearing supported shafts. A great attribute of their design often facilitates tighter running tolerances where the absolute minimum end clearance is needed. With rocker arms, both thrust washers and thrust bearings have been a Miller rocker original design specification since 1982 when NEEDLE BEARING THRUST WASHERS were implemented to control the excessive side thrust inherent of the **BOSS 429** Ford. To this date and the best of our knowledge, no other aftermarket rocker arm manufacturer uses them. Their use was also carried over to the existing NASCAR model Miller PRO-SHAFT™ systems, and offer the ultimate in reduced drag coefficients to this lateral side load that OEM side thrust pushrod geometry uses. All PRO-STAND™ systems use THRUST WASHERS on each side of the rocker body to prevent scuffing of aluminum-to-aluminum between the body and stand, even though this is over-engineering, as the side thrusts with most SB Chevy and Ford systems is far less than the Hemi designs from which their inspiration came.

TOP DEAD CENTER (TDC): Refers to the position of the PISTON (and crankshaft) in relation to the stroke of the crank. It is the ultimate top of the crankshaft's stroke, before initiating its travel back down.

TORSIONAL TWIST: Is the rotational flexing that occurs around an axis of any linear object where a force of torque has been imposed upon one end that is greater than the absorbed torque transferred through to the restrained or driven opposite end of said object. Camshafts experience this, so do crankshafts (to a lesser and different degree). With camshafts however, the effect is an actual retarding of timing between cam lobes at the opposite end of the cam from those near the driven end. Engine builders for many years have experimented with grinding the cam to anticipate this, by place the rear cylinder cam lobes 2, 3 or more degrees advanced from the lead cam lobes near the timing end of the cam, so when the engine is accelerated to peak operating RPMs where the torsional twist is at its highest, the now retarded rear cam lobes have been evened up with the lead cylinder lobes.

TORQUE: A force applied to or from a rotating axis; measured at a fixed distance from that axis. See: **HORSEPOWER.** (See: **NOTE** on p.139 under **PISTON VELOCITY**.)

TRACKING VELOCITY: Is the accuracy of how well the rocker arm converts linear information into radial motion from the cam. The closer to "in-line" that tangents remain with the linear path of the pushrod, the less deviation occurs from the cam's acceleration dynamics. Tracking velocity can be altered by placing the 90 degree tangent either closer to open valve or closed valve, with reverse consequences to each; but doing either alters the cam's profile at the valve, diluting or accelerating the valve's response of the cam information.

TRUNNION: This refers to the singular steel "shaft" within the rocker arm of a "stud" mounted design, having 2 bearings, 1 on each end, with a hole in its middle to accommodate the cylinder head's rocker mounting stud. Because a trunnion is *also* a **SHAFT**, we choose to call the other rocker arm design "*STAND SYSTEMS*" rather than "shaft systems."

> **NOTE:** *All MEI or PVS trunnions used were designed to be hollow, or what I called rifle-drilled, or gun-drilled, which is merely a euphemism for precision drilled. Solid trunnions are weaker, because the hole down an axis provides a second "wall" to resist breaking and for the hardening from heat treat to penetrate. This engineering technique is most advantageous to resist "torsional twist," such as axles but otherwise solid shafts absorbing a tensile or compressive load benefit too. We always machined a flat on both sides of the trunnion's mounting stud hole to remove the sharp sheer point where fractures begin.*

TUNGSTEN INERT GAS (TIG): A term for the process of welding, usually high alloy, non-ferrous or exotic metals, whereby the heat producing arc is electronically conveyed through an electrode made of tungsten, and the arc is protected from the oxygen atmosphere by an inert gas, such as Argon. Filler rod is usually required to be added to the parent metals being welded. A more production oriented version of this same welding principle is done with a welding machine that feeds the filler rod through the Inert Gas nozzle (often like a hand held gun). Known as **MIG** for Metal Inert Gas, the arc is conveyed by the power fed filler rod being consumed in the welding process, rather than an tungsten electrode, thus saving the more time consuming hand fed filler rod of the typical TIG process.

TURBINE INLET TEMPERATURE (TIT) (Turbocharger)**:** The critical temperature to know on turbocharged engines seeking maximum performance with minimal risk of overheating the **Turbocharger**, particularly on engines whereby the fuel to air ratio is adjustable. Measured by a temperature probe usually mounted within (but not into) two inches of the turbocharger's **Hot Section** inlet, where the collective primary tubes of the engine come together, it is a commonly accepted operating formula for the "sustained" maximum temperature to never exceed 1,650° Fahrenheit, and in most conservative circles of use, this limit is held to: 1,550° Fahrenheit. It should noted that these are collective flame/gas temperatures within the exhaust system, not component temperatures of the system, which run several hundred degrees less. However, the cast iron hot section of turbochargers operating in these upper temperature limits will literally become a glowing orange/red, exerting immense radiant heat to all components around them, especially in confined engine bay areas, where ventilation, space and heat shielding have not been implemented as part of the installation. Another important, but less noted (and used) term refers to temperatures taken on the discharge side of the Turbine section, known as **TOT** (below).

> **NOTE:** *These temperature limits <u>should not be confused</u> with **Diesel** engine designs, which operate much lower, nor should it be confused with **Exhaust Gas Temperature (EGT)**, which will usually be held to no more than 1,450° Fahrenheit, and measured by carefully mounted probes that are approximately 4" out from the cylinder head's exhaust port flange to avoid direct contact with flame front core. <u>DO NOT CONFUSE</u> **EGT** with **Turbine Outlet Temperature (TOT)**.*

TURBINE OUTLET TEMPERATURE (TOT) (Turbocharger)**:** This is the *exhaust gas temperature* exiting the Turbocharger after driving the Turbine, which is then passed through a connecting exhaust pipe and muffler on engines so equipped. As with the **TIT**, the temperature probe is mounted within 2" (two inches) of the Turbocharger's **Hot Section** outlet. This probe location, which sees the Turbo's cumulative temperature from operation under all types of loads, is instrumental in determining cool down time before shutdown of the engine; which is often required with most applications that have been operating at a high load, in preventing premature wear and failure. This cool down time can be anywhere from one to five minutes (or more) of low or no load operation before killing engine. This brief time is often determined by various manufacturers of so equipped vehicles, and it is all time required to lower the **TOT** temperature (which is sometimes called the **EGT**, because it may be the *only* temperature probe on the engine's exhaust) by 250° to 350° Fahrenheit, making all the difference in turbo life when it has been taken to these higher extremes under load.

TURBOCHARGER (left) (a.k.a.: "Turbo"): *In its simplest definition, turbo-charging is the process of forcing air into the engine's induction system through use of the engine's exhaust pressure to drive the intake turbine.*

Turbochargers are essentially two carefully designed and independent circular air chambers that spiral around impellers on each end of a common shaft; of which each are designed for entirely different tasks.

The first is a precisely balanced **Turbine Wheel** driven by the exhaust gases at very high speeds which in turn drives the **Compressor Wheel** that is designed with a very complex geometry for two stages of operation, that in turn forces fresh air into the engine's induction system.

The first stage is dictated by a smaller (minor) diameter group of blades known as the **INDUCER**, which grab and pull the incoming air drawn from the air cleaner (or an ambient intake source), which then "compresses" the air and raises its temperature by 100° or more, then forces it out through a second stage of blades with a larger (major) diameter, known as the **EXDUCER**, that force the pressurized air through a precisely shaped, spiraling passage in the aluminum "compressor" housing, also referred to as the **Cool Section**. The compressed air is [usually] then sent through an aluminum "heat exchanger" known as an **After-cooler** (often incorrectly referred to as an "intercooler") where the air temperature is reduced before being sent on to the engine's induction. Both sections, the hot and the cool, have inlets and outlets; and in addition, both have a means to control the pressure development within each.

On the Compressor side (Induction), a **Bypass Valve** is used, also known as an Over-boost Valve, which allows excessive pressure in the intake tract to escape when a predetermined value has been reached. On the Turbine side (Exhaust), a bypass section designed integrally with the cast iron housing, or separately connected to the discharge side of the hot section, known as a **Waste-gate**, provides excessive pressure from the engine's exhaust system from running the Turbine faster than necessary for the loads which it is designed for. Literally, the inlet side of the Turbine is connected through this alternate passage to the outlet side, and thus "bypasses" the Turbine itself, dumping directly into the final exhaust system. Some waste-gates are adjustable, either manually, or automatically, to compensate for different temperature and/or atmospheric pressure conditions.

TURBOCHARGER COMPRESSOR WHEEL (right): The carefully designed impeller of the **Turbocharger's** cool section side, known as the compressor. The compressor wheel has two distinctly separate designed stages to its blade geometry, whereby the inlet side of the compressor housing feeds inlet blades known as the **INDUCER**, which takes the incoming air, compresses it and directs it through the integral exit blades, known as the **EXDUCER**.

TURBORCHARGER COOL SECTION (left): This is the Aluminum housing half of the Turbocharger assembly, which contains the Compressor Wheel and accommodates the pressurizing of normal outside air into the boosted higher pressure air fed into the engine. It's called the Cool Section because although it is attached to the other half of the Turbocharger in an integral but separable way, by comparison to the often glowing red 1,500° plus temperatures of the **Hot Section**, it operates at a temperature only one-fifth of that. To improve power that hot air is cooled by an optional "*after-cooler*" before being forced into the injector's throttle body. **See: TURBOCHARGER.**

TURBOCHARGER TURBINE WHEEL (right): The carefully designed impeller of the Turbo's **Hot Section** which is the driving first stage of the **Turbocharger's** operation. The Turbine Wheel, which is driven at super-high RPMs of up to and over 200,000, often operates from sustained exhaust gas temperatures ranging between 1,400° to 1,650° Fahrenheit, but rarely more without incurring imminent damage. This temperature is measured as **TIT**, for "**Turbine Inlet Temperature**," whereby a probe is usually mounted within two inches of the Turbocharger's hot section inlet.

TURBOCHARGER HOT SECTION (left): The cast iron ***Hot Section*** flange on the right connects directly to the exhaust primary tube collector (or cylinder head's exhaust manifold) receiving high pressure, high temperature exhaust gases that drive the TURBINE WHEEL (above), which exits the center port that is connected to the remaining exhaust. Under normal operation this cast iron manifold reaches a cherry red glow that requires a heat shield "blanket" or thermal coating to minimize the engine compartment heat.

ADVANCED ROCKER ARM TECH

Jim Miller

TURBULENCE (Cylinder Head): Refers to uncontrolled air flow patterns and their dynamics. Specifically, it is the point at which **LAMINAR** airflow patterns leave controlled, smooth layer (or swirl) flow characteristics and collide in uncontrolled, random (and violent) shifting patterns, capable of motion in all directions, including "up-stream." Some poorly shaped ports inhibit these bad dynamics throughout the majority of their valve lift ranges. Two of the largest culprits to such ports are **PORT WINDOWS** that are too straight across the **SHORT SIDE RADIUS**, creating a "ski ramp" effect, where airflow leaves the port floor as it approaches the valve bowl area, effectively having no flow across the near side perimeter of the valve head because air is literally skipping over it. The second is too big of a transition from a smaller **CROSS-SECTIONAL-AREA** upstream; where the flow slows down too quickly as it enters this larger area; gets in its own way and tumbles.

TWISTED ROCKERS: This is a term that we use to describe an abnormal operating condition of a rocker arm being turned, or "twisted" on its mounting axis from the correct alignment required to maintain its reciprocating arc true and in-line with the valve's centerline. This is most prevalent on STUD mounted rocker arms where the pushrod has been offset and the stud location has been shifted to the same direction, an incorrect flaw done too often by cylinder head manufacturers, forcing the rocker arm to now be rotated out of alignment with the cam's centerline, as is needed on in-line valve array cylinder heads that use roller tip rocker arms, where this error most often occurs. NOTE: On OEM style BALL Fulcrum mounted rocker arms this problem is nearly non-existent, because with a ball fulcrum, the rocker arm can shift its twist sideways to keep the rocker's valve tip end flush atop the valve, even though it is twisted on the stud. The ball allows it to self align on two axis, where the needle bearings of a trunnion design roller rocker can only control one axis; and their axis must ALWAYS be in parallel alignment with the cam centerline on in-line valve array engines. Period. (See: p.52)

UNDER-ARCING: A term we use to describe the radial path a rocker arm makes on either end which reaches a 90° relationship to its axis with the valve centerline BEFORE reaching the MID-LIFT point of total sweep; causing it to cut beneath itself as it approaches peak lift. It is the opposite of **OVER-ARCING** in that the roller (or rocker tip) actually sweeps beneath itself as it moves toward full valve lift, a symptom that occurs from having a rocker pivot point that is too high.

VALVE BOWL (Cylinder Heads): Refers to the immediate material making up the inside diameter of the cylinder head port, just beneath the lowest angle of a valve seat, where the port is traditionally its smallest cross sectional area on a performance head; although many OEM heads, or stock type aftermarket replacement heads may in fact have even smaller ports further upstream (on the intake), to keep velocity high in near stock performance application engines. This area of space that is called the valve bowl is about 1/4" to 3/8" below the valve seat, depending on the head model. It is a very critical standard that works in harmony with the valve diameter to establish THE STANDARD for which a cylinder head's ultimate potential and efficiency are established; as well as the yard stick for port dimension limits that should not be exceeded in pursuit of high efficiency laminar flow ports, on any racing engine. The valve bowl and the valve are GROUND ZERO for all port dimensions and limits. **See: CROSS-SECTIONAL-AREA** (p.125).

VALVE CENTERLINE: In referring to rocker arm geometry, the valve's centerline is the most important angle to reference from. This is what must move with the least amount of side load, friction, or wasted motion. Additionally, it is the bearer of the valve spring. And when the various levels of motion, heat and impact are added together; the last thing that is needed – is to ask the valve to straighten itself against an over-arcing rocker arm that is throwing its side loads down at 80 cycles per second. The VALVE MUST move STRAIGHT up and down in its guide, and nothing else. The valve centerline is the basis for assuring the rocker arm operates accurately.

ADVANCED ROCKER ARM TECH

Jim Miller

VALVE CLEARANCE: Is a term that refers to **PISTON to VALVE (PV)**. A critical dimension, that is varied in some regards due to the engine application being built for (i.e., Drag Racing, Circle Track, Street, etc.), as well as the RPMs, camshaft type (Roller Tappet, Flat Tappet, Hydraulic Tappet), and also on the engine builder's preference. Some engine builders like more clearance than others, some don't have many options for how much clearance they can build for, due to cylinder head, compression ratio, valve lift and a few other factors that are more important than a certain safety value that exceeds what they may be confident they can "get by with." But it is a dangerous game. The published "norm" for many years has been .080" for the Intake valve and .100" or more for the Exhaust valve. The BOSS 429 Ford Hemi engine requires even more, by as much as .020" on each; while some high RPM small block engines using Titanium valves and reasonably high, but not insane valve lifts can get by with as little as .060" and .080", respectively. If you have the luxury of dictating what you need, without losing compression ratio from over cutting the pistons to accommodate this, then by all means run the higher figures. If you accidentally over-rev the engine, then this extra margin will have paid for itself.

> **NOTE:** *Piston to Valve clearance occurs during the **OVERLAP** cycle, when the piston is rising and chasing the Exhaust Valve UP as it is closing, while at the same time the INTAKE Valve is opening and coming down to meet the rising piston. Piston to Valve clearance is always affected by advancing or retarding the cam, and you can gain or lose accordingly. When changing to MID-LIFT rocker arms, you will also see more **"AREA-UNDER-THE-CURVE"** (quicker valve response at less crank position), so it is easy to lose Piston to Valve clearance here, if you're already at the minimum. **Another key point:** Do NOT make a parallel between increased VALVE LIFT and LOST Piston to Valve clearance. Because **PV** clearance happens at Overlap, it is not uncommon to see more valve lift have a minimal effect on valve clearance, anymore than it is not uncommon to LOSE Piston to Valve clearance changing to a cam with LESS valve lift. It all revolves around "timing" – and the PEAK valve lift can often have little effect to PV clearance.*

VALVE FACE (Cylinder Heads): Refers to the precision ground surface of the valve that actually seals against the mating surface of a like angle ground into the cylinder head, known as the **VALVE SEAT**. Usually, there is a 1/2 degree "positive" angle of interference between these two mating surfaces, meaning that the valve's outer edge contacts the seat first, to allow for valve head flex and efficiency of the valve's perimeter breathing (maximum diameter being used).

VALVE HEAD: The actual major diameter of the valve which encompasses the valve face angles used to seal against the valve seat of the cylinder head.

VALVE LASH: Is the mechanical clearance of free play between the valve tip and the rocker arm, designed to provide adequate clearance in keeping the valve's closing un-interfered with from operational heat and dynamics. Traditionally applied in meaning to mechanical cams, such as solid roller and solid flat tappet designs, it can technically be applied to the hydraulic cam followers too, even though this lash is controlled by hydraulic valve compression within the tappet, through metered bleed-off of the engine's oil pressure. For purposes of continuity across cam applications, we always use ZERO LASH in calculating **ROCKER RATIOS**, which you can learn more about on p.93.

NOTE: My **LASH RECOMMENDATIONS** are "**tight**." Always has been. Long before it became reduced to the lash recommendations of some cam companies. I would like to say that we pioneered it, but we didn't. My first introduction to tight lash, was in 1970 with the ***BOSS 429***, after falling victim to the published rhetoric of the factory's 1960s lash thinking of .020" to .025" on the Ford 427 FE, from two years prior. The lash settings for the BOSS 429 recommended **.013"** intake and **.015"** exhaust. This could have been argued against by older logic, noting that aluminum heads "expand more," so they need more lash. Motorcycles, my other love, routinely set lash by *rotating the pushrod freely*! That's tight! I recommend valve lash for the **INTAKE: .012" to .015",** and the **EXHAUST: .015" to .018"** for ANY mechanical cam. Especially if the cam companies recommend more! The tighter lash softens shock, adds a few degrees of duration (which most people can use, because cam companies routinely cut this short); and lastly will require setting less often, if you've done it correctly. *If you need more lash, you've got the wrong cam.*

VALVE LENGTH (Cylinder Heads): Is an easy to understand, self descriptive term with a couple of notes that should be kept in mind, especially for novice engine builders. First of all, in buying aftermarket valves, these are often referred to by an extended length over stock. So you'll often hear "hundred long," or " two hundred long" meaning .100" or .200" longer than stock, respectively. But be careful, because on some heads models of some manufacturers, these "stock lengths" changed over the years; so this can be a moving target if you're not careful. It's better to have a way of measuring what you really need, rather than just fitting pieces together because they are the most commonly used, or in stock. Too many times, engine building comes down to compromises on doing it right, because you're faced with buying parts that had a domino effect which forced you to buy related parts that weren't ideal. Most of the better valve manufacturers easily provide the **OAL** (Over All Length) for their various part numbers so you pick and choose with more reference to what you really need. If you don't find this information in their catalog, then just call their technical staff and ask. If they don't know, find another manufacturer! Secondly, all valves are measured and sold by this OAL, which includes the **VALVE MARGIN**; and this may be a different spec between some valve models and/or manufacturers. So be sure to ask about this when buying, and make sure you don't lose .020" from specific height you are looking for that is measured from the valve seat – which is what really matters. With the **OEM** of cylinder heads (Ford, Chevy, Dodge, etc.), this is known as the **GAUGE HEIGHT** for the valve seat.

> **NOTE:** Choosing the right valve length is very important to rocker arm geometry. There is no standard among cylinder head manufacturers for a height of the **VALVE TIP** above the SPRING PAD – so placing the roller path in the middle of the valve is not always feasible. The spring pad is often a taboo surface to mess with, and always the foundation for determining valve spring heights, valve spring retainer designs and the valve spring's *installed height*, which is necessary for adequate *coil bind clearance*. So it is only logical to use the spring pad for designing specific valve tip heights. Regardless of which rocker brand you use the valve tip is critical as the starting point of leverage for setting the rocker arm's installed geometry. Some very popular rocker brands are engineered with no relative choice of where the valve tip is at. They require that engine builder just figure it out for themselves, and unfortunately valve lengths are ordered randomly; often too long, and the stand "**SHIMS**" are always excessive – especially after the engine builder has really figured out how to set their installed rocker geometry.

VALVE LIFT: Is self descriptive, as the amount of lift the valve operates with. What should be noted though, is that this term is used (and unfortunately trusted as accurate) on **CAM CARDS** or in cam sales, when in fact the reality of it is that it is often NOT accurate. These are more appropriately termed "THEORETICAL" Valve Lift, while an engine's *true* valve lift is called "NET" Valve Lift. Net Valve Lift is usually LESS than "advertised," but sometimes when improper rocker arms are chosen or incorrectly installed, can be MORE than expected (although rare). In either event, expensive errors in critical clearances, like Piston to Valve can result from "trusting" theoretical calculations. Always check. Valve lift is the result of a multiplied **RATIO** by the rocker arm of **CAM LIFT**. Never really standardized before MID-LIFT, "NET" valve lift at "**ZERO LASH**" is the best ground zero method to calculate net valve lift by, because it leaves the variables such as "lash" up to the engine builder in determining the engine's final valve lift specifications.

> **NOTE:** See **CHOOSING VALVE LIFT** (p.97).

VALVE LIFT DYNAMICS (VLD): This is a term we use to describe the actual, net, valve lift events, in the same way they are analyzed at the cam: LIFT, DURATION and VELOCITY. Rate of acceleration and deceleration is the more accurate description of "velocity" as we use it here. What is so critical about this perspective, even more so than the cam's events, is this takes the entire valve-train system into account, and gives us the magnified results of cam information as it is translated through the rocker arm. Any changes in rocker arm geometry, either **DESIGN GEOMETRY** or **INSTALLED GEOMETRY**, directly and significantly influence the **VLD**, and the **TRACKING VELOCITY** (p.151). It is through plotting the **VLD** on graph paper that **AREA-UNDER-THE-CURVE** is seen, quantified and analyzed for what is really happening to study **HORSEPOWER** data.

> **NOTE:** Tracing the valve motion directly to establish a given engine's VLD is crucial to properly understanding the end consequences of cam designs and rocker geometry. CHARTING this with graph paper, which illustrates the Area-Under-The-Curve clearly is necessary. There are TWO methods for doing this: (1) Use an even increment of CRANK degrees (horizontally on the graph to mark TIME) and write down the amount of VALVE LIFT for each increment you've chosen; such as each 5° of rotation; or (2) Use an even increment of VALVE LIFT, such as .005", and write down the degrees of crank rotation seen. Both illustrate the same results, but from opposite perspectives. The second method will clearly show gains and losses of crank rotation due to over-arcing rocker geometry, and it is a test that can be done on any engine, with any rocker system, by merely doing it twice and changing the pushrod lengths a reasonable amount between the two tests, such as .050" or .100". Following that, you won't have too much trouble understanding how important precise **DESIGN GEOMETRY** rocker arms become, as well as exact length pushrods for proper **INSTALLED GEOMETRY**.

VALVE LOCKS: **See: VALVE SPRING RETAINER** (p.158).

VALVE MARGIN: Is the width of material on the valve head's major diameter, which lays between the valve's bottom face and the valve seat face. It is very critical to valve longevity, to disperse (or absorb) heat in conducting it to the cylinder head's valve seat. It is also critical to air flow, mostly at low lifts, and especially the exhaust, but also has an effect at mid-range lifts, depending on the port, combustion chamber and application. It is sometimes used in conjunction with a radius on the valve's bottom (chamber side) to induce flow across the exhaust, but most testing has shown this not to be beneficial, and in fact opposite. The argument will continue. One thing that few people argue are the minimum widths before throwing the valves away, or re-cutting their outside diameter to recover a too thin dimension, and these specs are: INTAKE. .030", EXHAUST .050". Although some engine builders have got away with .020" INT. and .040" EXH., anything less is real risk of burning, and anything in these ranges has already had a detrimental effect on air flow across the seat. We never rebuilt a head that was less than .030" INT. and .050" EXH. New design specs were always .050" INT. and .085-.100" EXH. Just my personal preferences. (**See: VALVE TIP LENGTH** for illustration of margin's location.)

VALVE POCKET (Cylinder Heads): Refers to the area just beneath the valve seat area of the cylinder head. It is usually 10% less area in size than the valve itself is, on HEMI style heads, and 12-15% less for WEDGE heads. It is also the transition portion of the port's short and long side radius` to the valve seat's bottom angles.

VALVE SEAT (Cylinder Heads): Refers to the precision ground angle in the cylinder head that provides a seating surface for the valve's **VALVE FACE** to close upon. It is usually .040-.080" wide for INTAKES, and .080-.120" for EXHAUST valves. Different engine designs, applications, materials and "tricks" by various engine builders will vary these general dimensions in both directions.

VALVE SEAT INSERT (Cylinder Heads): Refers to the independent component which is separately installed into a cylinder head in providing the precision valve seating surface, necessary with all non-ferrous (aluminum or magnesium) cylinder heads, or any iron heads which required an alternate valve seating material of higher strength, higher operating temperatures or simply restoration from original valve seat damage. Valve seat inserts are made of a variety of materials, from low tech iron to high tech beryllium, Stellite*, and other high temperature alloys.* (*Stellite is a trademarked name of the Deloro Stellite Company, and part of the Kennametal group.)* With titanium valves, these extra hard nickel and cobalt alloys can't be used without beating the valve faces to death. Emphasizing the importance of choosing the proper material in working with different fuels and various alloy valves.

> **NOTE:** Valve seat inserts must be installed carefully. Typically agreed in the industry that with aluminum heads, an interference fit (press fit) of .007" to .008" between the valve seat insert's outside diameter and the head's valve seat insert "pocket's" inside diameter. The machined surface of both must be concentric within .001" and smooth. Lastly, the best way to install them is to "freeze shrink" the inserts with dry ice and alcohol, or liquid nitrogen (harder to come by and more dangerous to use around clumsy hands). *TRICK of the trade: REMOVING valve seat inserts is easy.* Using a **TIG** welding torch to heat them up by circling around them about three or four times will anneal them; making their press fit dissolve and they will literally fall out of the head. Be careful not to keep the TIG tip close to the insert and avoid excessive heat going into the head which can hurt its heat treat. Yes, aluminum heads have a heat treat.

VALVE SPRING RETAINER (Cylinder Heads): Nearly self descriptive, the valve spring retainer is a flat appearing round disc-like item that sits atop the valve spring and affixes itself to the valve stem by two, tapered, steel cylindrical halves, known as **VALVE LOCKS** or **KEEPERS.** As we all know, valve locks act as a door stop in that their two halves are compressed against the valve stem by a corresponding tapered hole in the center of the retainer, of which the upward pressure imparted by the valve spring forces the tapers to "grip" against a matching male/female groove between the keepers and a valve stem. Valve spring retainers from the OEM are made in steel, but aftermarket uses aluminum and titanium, and are substituted from OEM with any new valve spring package to fit larger springs and often to reduce weight. Two basic angles of LOCKS are most often used, 7° and 10°, the latter believed by many to better resist pulling through on radical cams and high spring pressures. *With this rudimentary definition out of the way, we can now make the point of FIT needed in a stable valve spring package.*

> **NOTE: FIT** between the valve spring RETAINER and the VALVE SPRING is critical. Valve springs dance all over the place in an operating engine at high RPMs. Keeping this dancing to a minimum is critical. Keeping their friction at a minimum is critical too. Yet, they are often sold and promoted with tight fitting inner and outer springs as if this is a virtue. It isn't. Valve springs are horsepower robbing heat generators. The "step" that is machined in the valve spring retainer needs to have a near "snap fit" where the spring has no free-play to dance around. Because of the many retainer and valve spring variations and combinations used to fit a wide variety of cam kits, often, compromises are made in matching retainers with spring packages. It is difficult to get the proper, tight fit needed between the steps of the valve spring retainer for ALL combinations. On DUAL SPRING and TRIPLE SPRING combinations, it isn't uncommon to find the outer spring fit the machined step, while the inner springs do not. Or worse, the opposite might be found: the inner spring may be held intact by the retainer's protruding valve lock extension, while the middle and outer springs are loose fitting. STEP HEIGHT is also important for each of the corresponding springs within the outer spring, as each of these will **COIL BIND** at a different height; and these must be checked individually with the retainer, then as an assembly; noting which coil binds first. Ideally, we liked to see the OUTER spring COIL BIND a few thousandths before the inners. But the step height of the pad where each of the inner springs sits against the valve spring retainer should be of adequate height to preload the corresponding spring when assembled to the outer spring's needed height, and NOT be the first to coil bind in doing so.

VALVE STEM: The longest diameter portion of the valve which rides within the cylinder head's valve guide and adjoins the valve tip with the valve head, also encompassing the keeper groove or grooves, to affix the valve spring and retainer by split locks. This is also a wearing surface through its continued operation, showing signs of this wear along its outside surface as "scuffing" or galling – in worse cases – from a variety of usage reasons. In valve-train efficiency, the lack of proper rocker arm geometry becomes evident on the valve stems and valve tips, first, before most other symptoms are seen.

VALVE TIP: Upper end opposite the valve HEAD, where direct contact from the rocker arm is made to open and control closing of the valve on OHV engines. The VALVE TIP is everything; EVERYTHING. It is from here that all true consequences of the rocker arm's geometry occurs; yet, engine builders indiscriminately add longer and longer valves, thinking all they need to do is to use longer pushrods to compensate. **See: VALVE TIP LENGTH** (p.50).

VALVE TIP ADJUSTMENT: With higher end STAND MOUNTED ROCKERS ("shaft systems") using a common shaft for both the Intake and the Exhaust rockers of a common cylinder or an entire bank of rockers for one head, the VALVE TIP is the only way and the BEST way to accurately adjust the rocker arm's pivot point in relation to the valve tip. Since Exhaust valve lifts will differ from Intake valve lifts more often than not, there is no easy way to adjust the rocker's shaft height of one and not the other, since they are joined. Yet they NEED to be set accurately. By varying the VALVE TIP LENGTHS exclusively for this purpose, you establish precise **INSTALLED GEOMETRY** without compromise. For example: Most Exhaust valve lifts will be less than the Intake, so for an engine with .750" Net Intake Valve Lift, and .700" Net Exhaust Valve Lift, the MID-LIFT point for the Exhaust Valve will be .025" higher (half the difference) than the Intake Valve's MID-LIFT point. So to RAISE the ROCKER SHAFT in relation to the VALVE TIP, *without raising the shaft*, your only option is to *LOWER* the Exhaust Valve Tip by .025" *in comparison to the Intake Valve Tip*. If you need a MINIMUM VALVE TIP LENGTH (as most engines do) for Valve Spring Retainer clearance, rocker body clearance or whatever, then whatever this needs to be must fit the shorter height valve tip of the two valves; in our example's case this would be the Exhaust Valve. So you would add the required surplus to the *INTAKE VALVE*. From this baseline, you would now adjust your STAND HEIGHT for exactly what is needed to set your shaft's position with the Intake Valve. Naturally on stand mounted systems that have independent shafts and stands per rocker arm, this has no meaning.

VALVE TIP LENGTH ("VTL"): refers to the dimension taken from the end of the valve (tip) to the nearest edge of the KEEPER groove. Usually either .250" or .310", this is critical for standardizing to the retainers and valve springs you choose.

VALVE TIP HEIGHT ("VTH"): This is a term we established long, long ago (in our cylinder head prepping days) to standardized the HEIGHT of the valve tip over the valve spring PAD. It is the best way we know of to keep valve length requirements consistent in designing rocker arms. This is necessary because all heads have either an incline TO or FROM the valve of the rocker arm mounting pad (or studs). Making any deviations in valve lengths impossible to establish an accurate ROCKER LENGTH that keeps the roller in the middle of the valve tip. **See: VALVE TIP LENGTH** (p.50).

ADVANCED ROCKER ARM TECH

Jim Miller

VALVE-TRAIN: is the most temperamental and demanding system of an Over-Head Valve engine. Most engine failures are the result of something breaking in the valve-train, usually a valve spring or rocker arm. But most improvements in engine performance come from valve-train development. The most precisely designed and driving component of the valve-train, which is aptly referred to as the "heart" of the engine, is precision ground to tolerances less than 2/10 thousandths of an inch (.0002"), and computer designed to tolerances 10 times greater. This precision component works in harmony with three other components of the valve-train to open the valve, and determines all of the critical timing for the engine's ability to breathe the greatest amount of airflow, while venting wasted gases, all in an effort to produce maximum horsepower. It is known as the ***CAMSHAFT***.

VARIABLE RATIO (a.k.a.: ACCELERATED RATIO): This is a confusing "hype" term which is easily mistaken as a virtue for what is really just an over-arcing rocker arm, for which the performance of has no bearing in fact. Mainly because there is no virtue to a rocker arm exceeding its arc more than necessary to perform its linear translation of cam information to the valve. But "variable ratio" and "quick lift" terms are really promoting a rocker arm which has a different (higher) ratio in the closed valve position, which then decreases in ratio as it approaches full lift position. Let me explain (again):

> Technically, all rocker arms have a "variable ratio." Rocker arms go through an exchange of virtual "contact" point lengths in their center-to-center dimensions during operation, because the lengths from the axis on both sides of the rocker arm are relative to a straight line dimension which is measured through its axis at a right angle to each: the valve and pushrod. Since these rocker arm ends are rotating around an axis, when measuring them relative to the in-line motions of the valve and pushrod, their contact lengths actually change with this arc; even though the actual center-to-center of these tangent points within the rocker body are fixed. Technically, this varies their ratios a "small" amount throughout their lift cycle, but *only* when MID-LIFT geometry is utilized. MID-LIFT minimizes this to a predefined, predictable standard which guarantees the *ultimate minimum* in-and-out motion on BOTH sides of the rocker arm. If different geometry is used, where tangent points are set for 90° at any valve lift or cam lift other than MID-LIFT, then these ratio changes are greater as they move through the arcing process. Since "ratio" is only measured at FULL LIFT, a rocker manufacturer can hide this excessive (wasteful) motion by setting the center-to-center dimensions so that their publicized ratio is close at full lift, then attach trick sounding terms to this wasteful and inefficient design, and bombard engine builders with lots of advertising that this is a virtue. It is not. Ratio, by definition, is only a snapshot of a specification which does not depict the real waste of cam acceleration and duration throughout the more critical stages of valve lift, even though net valve lift (and ratio) may be close to what is advertised.
>
> To see the losses in valve acceleration from the extra crank degree rotation (**area-under-the-curve**), one needs to map the valve lift cycle in fine increments from closed to full open, then closed again, by measuring net valve lift directly at the valve and noting on graph paper the required degrees of crank rotation for each .010" or .020" (or whatever you choose) of valve lift to measure by. *You can also do the test from the opposite perspective, and pick a specific crank degree step, like every 5 degrees (something small), and jot down the valve lift on graph paper. This latter example gives you a valve reaction perspective which will show lost valve lift at respective crank positions between one geometry test and another; while the former example gives you a cam reaction perspective, which illustrates the change of crank degrees for a given valve lift.* You can do this test with two different pushrod lengths, back to back, of just .050" difference. Doesn't matter if you're too long or too short, just the results from the change should get your attention. *You'll soon see for yourself that the closer you get to a true MID-LIFT geometry, the more quickly the valve will react to each degree of crank rotation.*

Understand what is going on here: **Rocker ratio** is a mechanical division of the pushrod end into the roller tip (or shoe tip) end of the rocker. If you want to increase ratio, you shorten the pushrod dimension to the axis, or lengthen the roller tip dimension to the axis. But on rocker designs marketed under misleading terms there is no shifting or moving parts where the pushrod is at, nor does the roller tip move in and out within the rocker body to effect a different physical dimension to the axis (although some creative European engineers have patented some interesting concepts along these lines, they don't apply to American **OHV** engines). So how can the rocker arm have variable ratios, or accelerating ratios, or quick lifting ratios? It simply has to do with the angle of the pivot point (shaft/trunnion axis) with the axis of the pushrod tip and roller tip. The angle when the greatest response velocity to cam motion will occur is when there is a 90° relationship to whichever side of the rocker you are trying to see peak velocity, remembering that the roller tip end will only reach whatever the pushrod end dictates by its peak 90° angle and the cam. By setting the installed geometry at this 90° angle in the CLOSED valve position, the instant off the seat velocity is "accelerated" in comparison to the other lift points in the valve lift cycle. In fact, they will be decreasingly slower from that point on. In addition, the natural laws of physics will dictate that the arc from this starting point is increased proportionately, often doubling or tripling (or more) the in-and-out sweep.

Of course the other byproduct to this design is increased side loads on all contact points which endure this increased angle of attack; the valve stem in the valve guide, the pushrod tip into the adjusting screw or cup (often breaking them more easily), and let's not forget the additional side loads imposed on the bottom of the pushrod, where the resisting forces are now loading into the **cam follower** at a greater angle, and whip harmonics, which increases loads and wear, and drag coefficients within the tappet bore. None of this even touches the wasted cam information (mentioned above) through excessive motion in-and-out. Had enough? As noted elsewhere here, when you raise the rocker arm's pivot points to shift the geometry for promoting increased velocity off the seat, the rocker arm reverses the process as it closes. So if you've got a rocker design that sets geometry so that you've lifted the valve off the seat quickly, which inevitably slows down moving toward full lift, then when it has to close the valve will leave the full open position slowly and accelerate faster toward the closed position (i.e., the valve seat). *This is the most destructive scenario you could wish for.* You want to always set the valve down ON THE SEAT as gently as possible to minimize valve bounce. This is "*Engine building 101*." One term they've coined for this is called their "quick lift" technology; which is nothing more than "over-arcing" by another name. Don't be fooled! To read more details of these symptoms: **See: OVER-ARCING** (p.137); **UNDER-ARCING** (p.154); and **ROCKER RATIOS** (p.93).

VELOCITY (Generally): Means the *rate of speed* of any vehicle, item, component, fluid (including air) as measured in some constant value of time calculated by distance; or distance calculated by time. *Angular Velocity* is another derivative sometimes used in rocker arm analysis with comparisons of rocker arm lengths and their affect with valve lifts. ***PORT VELOCITY***, ***CAM VELOCITY*** and ***PISTON VELOCITY*** are all common terms used with engine builders.

VOLUMETRIC EFFICIENCY (VE): Is a percentage measurement of NET cylinder filling and operational throughput of airflow through the engine in comparison to its displacement. More simply stated, a cylinder having a volume of 100 cubic inches should theoretically be able to only hold 100 cubic inches. In so doing, it has achieved 100% "volumetric efficiency." In reality though, engines utilizing **RAM TUNING** for specific RPM ranges can optimize their efficiency beyond 100%, going over 110% or more. This is especially impressive when you keep in mind that displacement values of a cylinder are limited to the actual cam timing events of when the valves are opened and closed in relation to piston position. Since the valves are opening and closing before and after the piston's ultimate TOP and BOTTOM positions to *anticipate* delays in response of air flow momentum, inertia and response time of the **FLOW DEPRESSION** traveling up the manifold runners that is created by the pistons, this superficial perspective of filling 100 cubic inches for 100% **VE** isn't as easy as implied, and the comparison isn't quite fair. **See: NECP** (p.135) and **COMPRESSION RATIO** (p.124).

> **NOTE:** The tuning tricks mentioned here for optimizing **VE** applies to **NATURALLY ASPIRATED** engines, of peak high performance design utilizing maximum effort components with peak RPMs being reached. Supercharged, Turbo-charged and Nitrous-oxide induced systems realize over-charged **VE** through unnatural pressurizing of the cylinder.

WASTEGATE (Turbocharger): The secondary passageway of a **Turbocharger's** exhaust driven hot section side, that provides either a fixed or adjustable relief of the engine's exhaust system pressure around the Turbocharger's **Turbine**, to limit unnecessary speed and Turbo pressure beyond a specifically designed amount for the engine in question.

WET FLOW (Cylinder Heads): Refers to the relative air flow characteristics of a cylinder head's intake port in the more relevant "wet" state that an air/fuel mixture would have, in comparison to the "**Dry Flow**" characteristics that most flow bench study is done under, where only the dry, ambient air pulled through the port by the flow bench, and used to develop port shape and flow quality values. Although used for many years with specialized flow bench modifications or dedicated designs, it adds another dynamic that can not really be duplicated in a laboratory test stand condition, as it really acts in operation within a competitive engine. Violent inertia pulsations of the heavier liquid within the air flow jet stream from reversion waves; rapid temperature variations, port surface texture effect at these different speeds within the port, in most cases open up more questions to a comparison of dry flow analysis than what may get answered. Which is why most port development to this day is still constrained to **dry flow** testing, backed up by dyno testing, where more real world data acquisition can be taken and correlated to dry flow data.

WEAR PATTERN (a.k.a. *ROLL PATTERN and WITNESS MARK*): The actual wear pattern made upon the valve stem tip. Very predictable with MID-LIFT geometry, but too often used to set geometry which is NOT an accurate method because its accuracy is taken at a right angle to its results. Unfortunately, numerous tools are based on this measurement which lends a false sense of security.

~jM

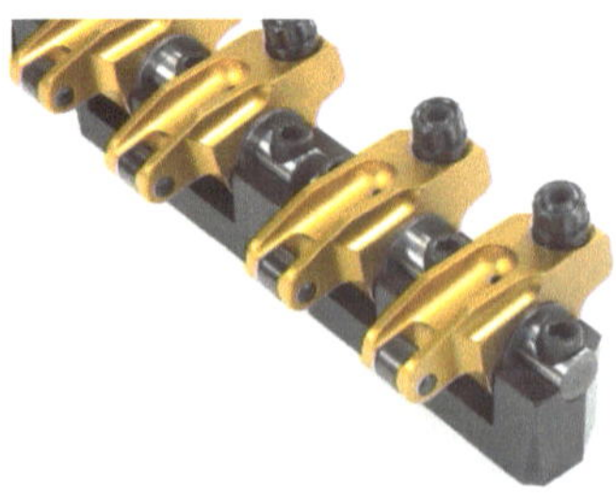

HISTORY OF MID-LIFT™

HOW MID-LIFT™ BEGAN

Beginning in the 1960s companies designed their rocker arms in the "closed valve" position with their priority to only fit the rocker arm to the head, *not the engine*, thus disregarding the pushrod's angle of influence with the valve.

Jim Miller understood the pushrod end of rocker motion needed to follow the cam precisely, and to do this it needed to have the least amount of in-and-out motion. Indiscriminately raising and lower the rocker arm, or the adjusting screw, defeated this.

The same principle applied to the valve side of the rocker. So Jim designed the rocker to be 90° with the valve and pushrod simultaneously at "Mid-Lift" of the total "NET" valve lift.

This influenced all of the critical dimensions, as it equally divided the rocker arm's circular motion (arc) on BOTH sides, which Jim emphasized was critical to not only minimizing back-and-forth sweep atop the valve, but also providing the maximum information of cam dynamics to the rocker arm with the ultimate least amount of wasted motion. This wasted additional motion that could be found in all previous rocker manufacturers required the crank to turn several additional degrees of rotation to effect the same amount of valve lift through the rocker arm. Miller's MID-LIFT™ principle of design set in stone a formula that would apply to any valve lift, provided the engine builder set the "installed geometry" for the net valve lift of the engine, a mandatory task with any rocker. Finally, a standard which determined and predicted every other criteria needed throughout the valve-train for optimum efficiency! In 1974, the phrase Jim kept using was coined: "Mid-Lift" which became the familiar terminology to his 1978 Patent application, filed two years later in 1980, and issued ***December 28, 1982***.

4365785

THE UNITED STATES OF AMERICA

TO ALL TO WHOM THESE PRESENTS SHALL COME:

Whereas, THERE HAS BEEN PRESENTED TO THE Commissioner of Patents and Trademarks A PETITION PRAYING FOR THE GRANT OF LETTERS PATENT FOR AN ALLEGED NEW AND USEFUL INVENTION THE TITLE AND DESCRIPTION OF WHICH ARE CONTAINED IN THE SPECIFICATION OF WHICH A COPY IS HEREUNTO ANNEXED AND MADE A PART HEREOF, AND THE VARIOUS REQUIREMENTS OF LAW IN SUCH CASES MADE AND PROVIDED HAVE BEEN COMPLIED WITH, AND THE TITLE THERETO IS, FROM THE RECORDS OF THE PATENT AND TRADEMARK OFFICE IN THE CLAIMANT(S) INDICATED IN THE SAID COPY, AND WHEREAS, UPON DUE EXAMINATION MADE, THE SAID CLAIMANT(S) IS (ARE) ADJUDGED TO BE ENTITLED TO A PATENT UNDER THE LAW.

NOW, THEREFORE, THESE Letters Patent ARE TO GRANT UNTO THE SAID CLAIMANT(S) AND THE SUCCESSORS, HEIRS OR ASSIGNS OF THE SAID CLAIMANT(S) FOR THE TERM OF SEVENTEEN YEARS FROM THE DATE OF THIS GRANT, SUBJECT TO THE PAYMENT OF ISSUE FEES AS PROVIDED BY LAW, THE RIGHT TO EXCLUDE OTHERS FROM MAKING, USING OR SELLING THE SAID INVENTION THROUGHOUT THE UNITED STATES.

In testimony whereof *I have hereunto set my hand and caused the seal of the* Patent and Trademark Office *to be affixed at the City of Washington this twenty-eighth day of December in the year of our Lord one thousand nine hundred and eighty-two, and of the Independence of the United States of America the two hundred and seventh.*

Attest:

Attesting Officer.

Commissioner of Patents and Trademarks.

The **MID-LIFT™ Patent: #4,365,785**.

HOW MID-LIFT™ BEGAN

United States Patent [19] [11] **4,365,785**

Miller [45] **Dec. 28, 1982**

[54] **ROCKER-ARM HAVING PERPENDICULAR GEOMETRY AT VALVE MID-LIFT**

[76] Inventor: **James M. Miller,** Galesburg, Ill. 61401

[21] Appl. No.: **211,638**

[22] Filed: **Dec. 1, 1980**

[51] **Int. Cl.**^{3} **F01L 1/18**; F16K 31/44

[52] **U.S. Cl.** **251/231**; 123/90.39; 123/90.45; 74/519; 74/522

[58] **Field of Search** 74/519, 522; 123/90.39, 123/90.41, 90.45; 251/231, 234

[56] **References Cited**

U.S. PATENT DOCUMENTS

3,066,659	12/1962	Colton	123/90.39
3,251,350	5/1966	Thompson	123/90.41
3,466,073	9/1969	Pohle	123/90.39
4,182,290	1/1980	Pohle	123/90.39
4,187,810	2/1980	Buehner	123/90.39

FOREIGN PATENT DOCUMENTS

666675	2/1952	United Kingdom	123/90.41

Primary Examiner—George L. Walton

[57] **ABSTRACT**

Roller-tip rocker-arm for high-performance internal combustion engines, pivotable about a trunnion therethrough, having a roller in the tip thereof for contacting the stem of a valve depressed (i.e., in the engine, per se, "lifted" from its seat in the "closed" position) thereby and having on the other end (i.e., tail) thereof an arcuate surface for contacting and abutting the cooperatively shaped end of a pushrod, the latter of which lifts said tail, wherein the roller and arcuate surface are so positioned so that, at the point in the motion of the rocker-arm wherein the valve stem is at a location corresponding to the mid-lift of the valve to which it is attached, the angles (a) that the axis of the valve stem defines with the axis intersecting the axes of the roller and trunnion and (b) that the axis of the pushrod defines with the axis intersecting the center of the arcuate surface and the axis of the trunnion are both essentially 90 degrees.

Such geometry eliminates the under-arcing and/or over-arcing of the roller-tip and/or pushrod characteristic of conventional rocker-arm operation and advantageously allows significantly improved engine performance.

Optional means are provided on the rocker-arm to allow maximum accuracy in adjusting the position of rocker-arm mounting in the engine.

16 Claims, 4 Drawing Figures

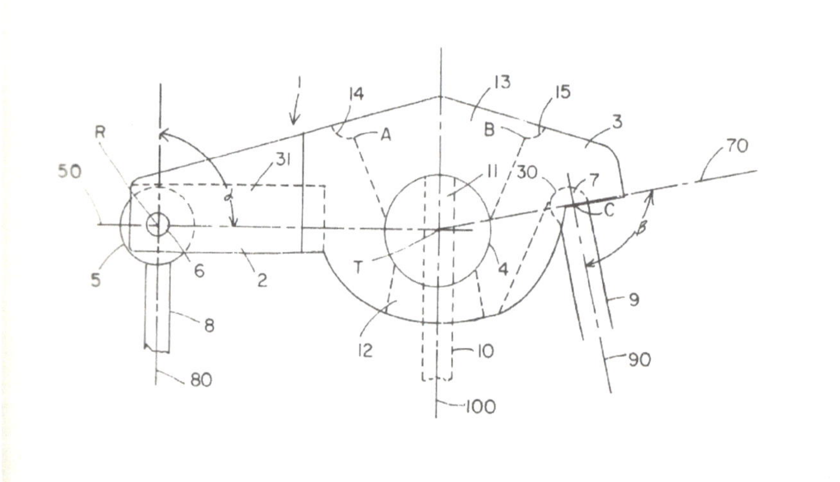

There are two kinds of Patents: (1) **Design**; and (2) ***Utility***. The utility Patent is the most preferred, most demanding, and requires the greatest proof of novelty to a specific sector (industry); and offers the greatest protection.

The MID-LIFT™ Patent is a utility Patent.

A Design Patent is more limited, easier to acquire, and has liberal criteria for acquisition and definition, and accordingly, has less strength for enforcing.

To validate a Patent, Jim's Patent attorney had to do a global search of prior Patents and claims; which proved no precedents existed for the simplicity of precision Jim specified for an OHV engine.

Jim wasn't satisfied with only knowing he developed something novel; he wanted to know "why?" How is it that something which seemed so *common sense* to him was overlooked by the entire racing industry, let alone Detroit's highly acclaimed engineers?

The search for this answer couldn't be done by any attorney. It could only be done by Jim; at least with regard to what existed at the time for roller tip rocker arms.

In 1974, the companies actually making roller tip rocker arms was a short list. Among them were Crane, Crower, Isky, Harland Sharp, and a couple less popular efforts. Dan Jesel was years from making his first rockers. Larry Torres (T&D) was also years away from copying Jesel to start his rocker arm endeavors. And the cheaper warehouse brands of the 1990s up through today didn't exist. The Chinese rockers that have discretely permeated the industry under a myriad of "private label" names was nearly 30 years away from their first foray into the otherwise "American made" only performance aftermarket.

HOW MID-LIFT™ BEGAN

THE QUESTION

So how could this be? How could an industry of multi-million dollar cam companies make such an oversight, let alone the MIT grads that filled the engineering rooms of Detroit's most competitive automakers? Finding the answer to this confounding question became necessary. Within a year the answer would be known, but two years was spent confirming the evolution of rocker variations across several decades which allowed this mistake to pass.

In 1958, *Harland Sharp*, an avid automotive enthusiast and innovative machinist, took a standard OEM stamped steel "shoe" type rocker arm and laid it on its side to trace its silhouette on a piece of paper.

Knowing he wanted to place a roller tip on the end, Harland traced a roughly .600" diameter roller, placing the bottom of the roller's diameter on the same horizon as the shoe's contact pad of the original rocker arm. This was the mistake.

Harland marked a center to this circle for the location of the roller pin, then using a scribe, he transferred his silhouette to a block of aluminum where he proceeded to machine his roller tip rocker body.

Several years later, Harvey Crane asked Harland to make rocker arms for his company, *CRANE CAMS*. However, within a short period of time, Crane's excellent national marketing outstripped Harland's production capability and forced CRANE to use other sources, eventually making them in-house. Having already established an identity with Harland's original roller rocker arm, CRANE simply continued with Harland's same basic design and the mistake was copied.

During the 1960s several of today's top cam companies began making their own line of rocker arms, and although many would no doubt argue their design was unique to their own ideas, the fact of the matter is that same Harland Sharp mistake continued to be copied throughout the industry.

RIGHT: The three generations of Crane stud mount SB rocker arms span more than thirty (30) years of evolution. The first model (left) inspired by Harland Sharp, illustrates how simple (and small) the original designs were. Despite the increased body mass, Harland's original *geometry* mistake remained; copied by Crower, Isky and others for decades.

HOW MID-LIFT™ BEGAN

THE ERROR

What was the error?...

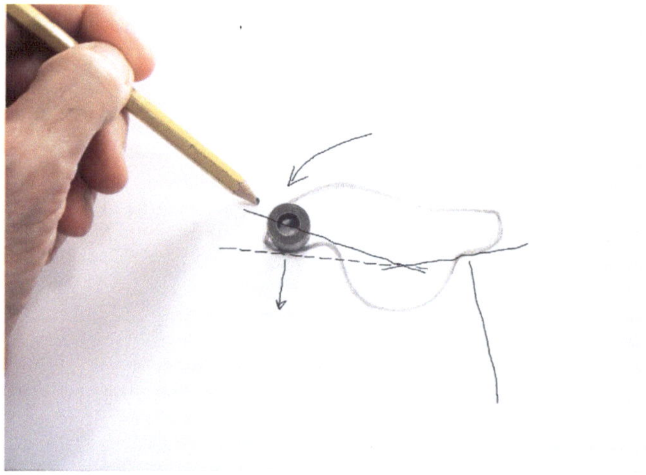

In much the same way the designing of a flat tappet cam is measured differently than the roller tappet cam, so too is the principle of design with a roller tip rocker arm. For on a "shoe" rocker arm, it is the bottom contact pad that has traditionally been used for angular measurement with the axis of the fulcrum (trunnion or shaft), but with a "roller tip" it is the axis of the roller that is used for angular motion. By placing the bottom of the roller on the same horizon as the shoe, the measurement was off by half the diameter of the .600" roller, or .300". In angular terms on the small block rocker arm, this works out to more than 12 degrees of error. To set this in perspective, the rocker arm only moves 24 degrees or so for a valve lift of .600" making the rocker arm wrong by 50% of its motion, regardless of the pushrod length or the rocker's height to the valve tip!

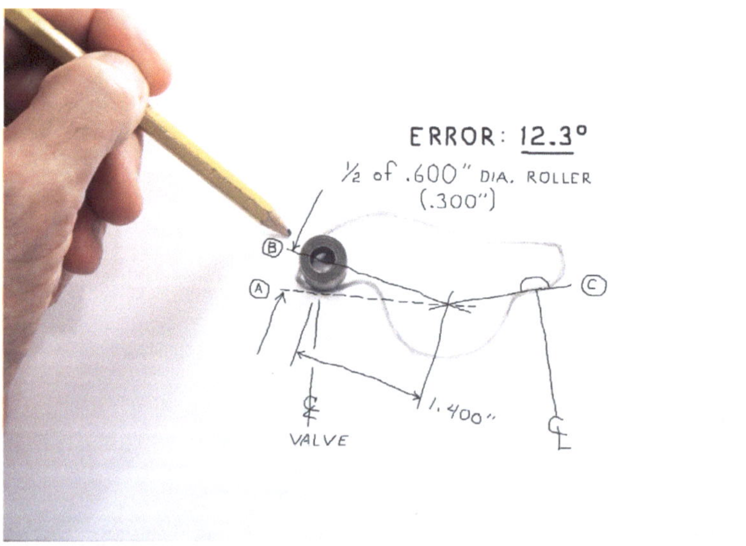

Line "A" represents the contact pad of the rocker, traditionally used for rocker design under the "**1/3 RULE.**" But with roller tappets or roller rockers it is the axis of the roller which matters, not the outer diameter, which is what Harland used, as shown. This moved the roller axis up by half of its .625" diameter. This extra .312" error had the effect of moving the push rod "cup" UP (shown as "C") by more than 12 degrees. The rocker arm only moves about 24 degrees, so no matter what pushrod length you choose, only one side of the rocker arm can be set for minimal back and forth motion, while the other end suffers. "B" represents this increased angle with "C" while the dashed line beneath "C" shows where MID-LIFT geometry should be.

NOTE: *The late Harland Sharp's background with racing interests dated back to the 1950s, when many of racing's largest and most well known companies were first beginning. The original hand written notes and drawings Harland sent the author during the 1970s are still kept as a reminder to the legacy of this history.*

HOW MID-LIFT™ EVOLVED

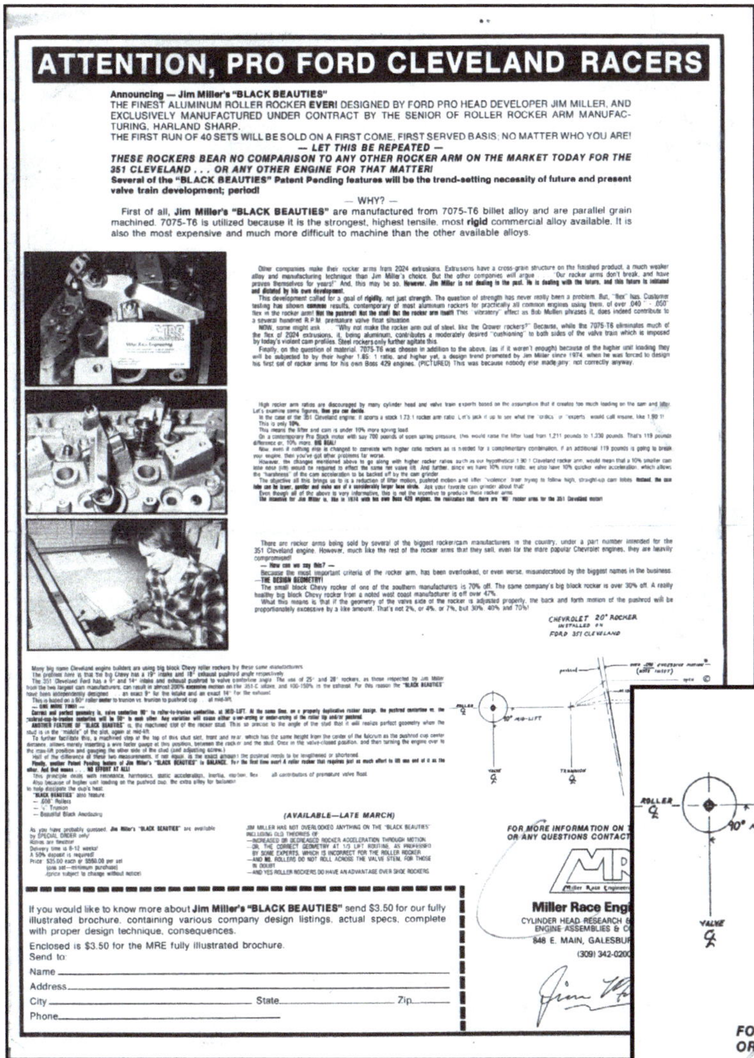

ATTENTION, PRO FORD CLEVELAND RACERS

Announcing — Jim Miller's "BLACK BEAUTIES"
THE FINEST ALUMINUM ROLLER ROCKER **EVER!** DESIGNED BY FORD PRO HEAD DEVELOPER JIM MILLER, AND EXCLUSIVELY MANUFACTURED UNDER CONTRACT BY THE SENIOR OF ROLLER ROCKER ARM MANUFACTURING, HARLAND SHARP.
THE FIRST RUN OF 40 SETS WILL BE SOLD ON A FIRST COME, FIRST SERVED BASIS; NO MATTER WHO YOU ARE!

— LET THIS BE REPEATED —

THESE ROCKERS BEAR NO COMPARISON TO ANY OTHER ROCKER ARM ON THE MARKET TODAY FOR THE 351 CLEVELAND . . . OR ANY OTHER ENGINE FOR THAT MATTER!
Several of the "BLACK BEAUTIES" Patent Pending features will be the trend-setting necessity of future and present valve train development; period!

— WHY? —

First of all, **Jim Miller's "BLACK BEAUTIES"** are manufactured from 7075-T6 billet alloy and are parallel grain machined. 7075-T6 is utilized because it is the strongest, highest tensile, most **rigid** commercial alloy available. It is also the most expensive and much more difficult to machine than the other available alloys.

Other companies make their rocker arms from 2024 extrusions. Extrusions have a cross-grain structure on the finished product, a much weaker alloy and manufacturing technique than Jim Miller's choice. But the other companies will argue . . . "Our rocker arms don't break, and have proven themselves for years!" And, this may be so. **However, Jim Miller is not dealing in the past. He is dealing with the future, and this future is initiated and dictated by his own development.**

This development called for a goal of **rigidity**, not just strength. The question of strength has never really been a problem. But, "flex" has. Customer testing has shown **common** results, contemporary of most aluminum rockers for practically all common engines using them, of over .040" - .050" flex in the rocker arm! **Not the pushrod! Not the stud! But the rocker arm itself!** This "vibratory" effect as Bob Mullen phrases it, does indeed contribute to a several hundred R.P.M. premature valve float situation.

NOW, some might ask . . . "Why not make the rocker arm out of steel, like the Crower rockers?" Because, while the 7075-T6 eliminates much of the flex of 2024 extrusions, it, being aluminum, contributes a moderately desired "cushioning" to both sides of the valve train which is imposed by today's violent cam profiles. Steel rockers only further agitate this.

CHEVROLET 20° ROCKER INSTALLED ON FORD 351 CLEVELAND

(AVAILABLE—LATE MARCH)

As you have probably guessed, **Jim Miller's "BLACK BEAUTIES"** are available by SPECIAL ORDER only!
Ratios are flexible!
Delivery time is 8-12 weeks!
A 50% deposit is required!
Price: $35.00 each or $550.00 per set
(one set—minimum purchase)
(price subject to change without notice)

JIM MILLER HAS NOT OVERLOOKED ANYTHING ON THE "BLACK BEAUTIES" INCLUDING OLD THEORIES OF:
—INCREASED OR DECREASED ROCKER ACCELERATION THROUGH MOTION.
—OR, THE CORRECT GEOMETRY AT 1/3 LIFT ROUTINE, AS PROFESSED BY SOME EXPERTS, WHICH IS INCORRECT FOR THE ROLLER ROCKER.
—AND NO, ROLLERS DO NOT ROLL ACROSS THE VALVE STEM, FOR THOSE IN DOUBT.
—AND YES ROLLER ROCKERS DO HAVE AN ADVANTAGE OVER SHOE ROCKERS.

If you would like to know more about **Jim Miller's "BLACK BEAUTIES"** send $3.50 for our fully illustrated brochure, containing various company design listings, actual specs, complete with proper design technique, consequences.

Enclosed is $3.50 for the MRE fully illustrated brochure.
Send to:
Name ______
Address ______
City ______ State ______ Zip ______
Phone ______

February 29, 1980 ***issue of National Dragster***, marked the public debut of **MID-LIFT** geometry to the racing industry.

This one time full page ad developed calls from teams all over the country; including companies like TRW, who connected the issues Jim raised to their problem of valve tip losses on their titanium valves being used by the Stock Car (NASCAR) teams, and elsewhere.

Bob Glidden, had already been "advised" on setting up MID-LIFT for the BOSS rockers Jim made for his de-stroked 370 inch BOSS Pro Stock, after NHRA over-taxed his record setting Cleveland with too much weight to compete.

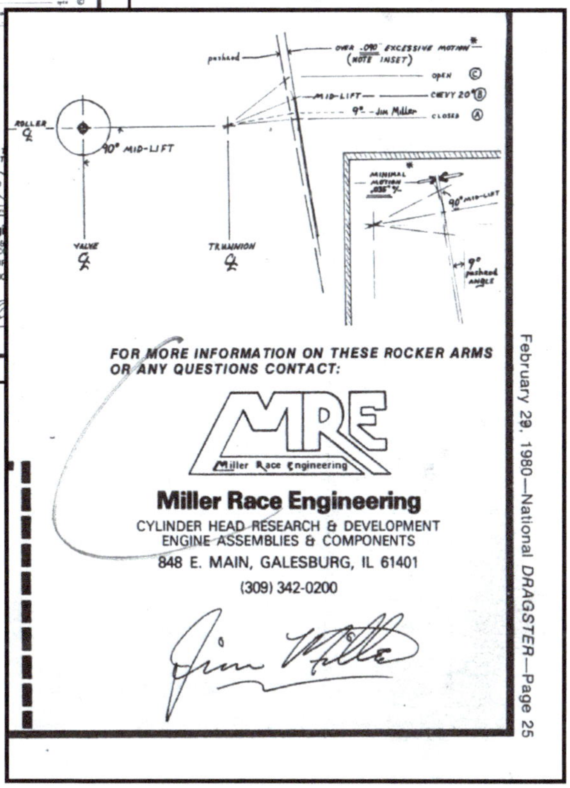

HOW MID-LIFT™ EVOLVED

PULLING POWER Magazine, **December 1984**, Jim Miller's now familiar illustrations debut.

FIGURE 2-A

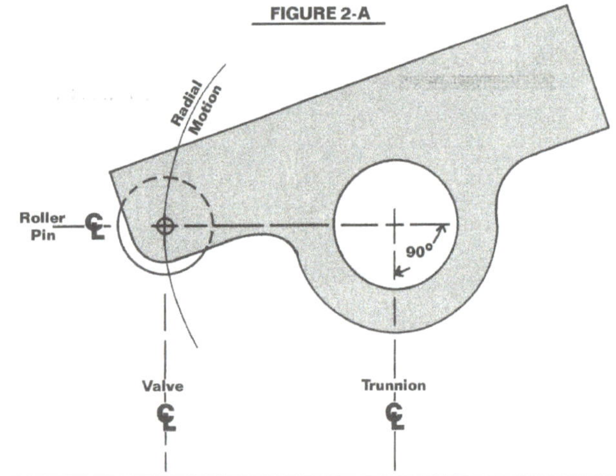

This view shows the rocker arm silhouette in the "mid-lift" position. (Or half of the *net* valve lift)

degree example. Since the rocker only moves a fraction of this amount, there is a wide variance of locations where the rocker arm can be mounted, which only may use one portion of these three variables. This is what happens most often. Most particularly, the rocker arm trunnion is too low, and the arcing motion is only outward. In some of the better cases, or perhaps I should say, the lesser of evils, may allow mostly an outward arc, with just a little bit of a downward motion as the valve approaches its full open position. (See illustrations.) The opposite of this problem, would be if the rocker arm was mounted too high, then one arcing motion would be biased toward an inward sweep.

You do not want any of these conditions. Even though the consequences might be less obvious, there is a price you're paying, and the blame may be attributed to that "misled direction of correction," such as too violent of a cam grind, or something else with similar symptoms.

Of this entire duration of the radial arc the rocker makes, there is only "ONE" time when this rocker arm is pushing down, in a direct line with the valve centerline. There is absolutely no side loading being applied at this time. And this is when the rocker arm is at a 90-degree position with the linear line of motion. If you have the rocker arm mounted too high on the stud, and the trunnion centerline and roller centerline are 90-degrees when the valve is closed, then common sense can only tell you that the rocker is going to make an inward sweep to open the valve. And just the opposite if the rocker arm is mounted too low. Additionally, the wear pattern on the valve stem tip is going to be its greatest under these above mentioned circumstances.

What you DO want, is the average of these circumstances. The rocker arm MUST execute ALL three of the radial points of motion. Outward . . . down . . . and inward.

And to do this, the rocker arm must be square at "Mid-lift." That is: the imaginary centerline of the rocker trunnion passing through the roller tip centerline, should be "square" or 90 degrees with the VALVE centerline, when the valve is half way open. Or at mid-lift. As can be seen, this gives you the ultimately smallest contact pattern. Now that you understand "why," I can show you the easiest and most accurate way to check this. But first, I better explain a couple of points I brought up earlier which may confuse you when you are trying to set this, or understand.

The reason I am so "down" on pushrod checkers, is that the third point of design, that being the pushrod cup, has different heights from one rocker arm brand to another. Consequently, the pushrod length required to adjust the rocker arm geometry would have to be different from brand to brand. Also, the pushrod checker has nothing to do with allowing for the NET valve lift to be applied. If you do not know what the net valve lift is going to be, there is no way in hell that you're going to set the geometry, as I've illustrated. Which brings me to another point. There is nobody, that I know of, which is as exact about setting the geometry as I am illustrating in this story. Even the World Champions I know, generalize the rocker position to achieve a minimal wear pattern only, and that is it. Many times, only getting themselves close to what is needed. This does not mean that this is not as critical as I've mentioned. It means that this is another point of engine assembly which you will find is a new "standard," much in the same way, as the use of a torque plate on a block prior to final honing. This blatant confusion of proper rocker arm design and application is further exemplified, by the variance of roller to trunnion center to centers. Not only from one manufacturer to another. But between the same line of rockers for the same application from the same manufacturer, offering different ratio's of that line (i.e. Small block Chevy 1.5-1 ratio, vs. 1.65-1 ratio). This is why you cannot even begin to use the rollers position on the valve stem, as a guideline to proper geometry. As long as the wear pattern is being fully contacted to the tip of the valve, and the "Mid-lift" principle of geometry is employed, it makes no difference at all if the roller is biased to one half of the stem tip. Additionally, you'll notice that on practically all of the American V8 engines relating to this article, the rocker arm STUD is inclined several degrees towards the valve centerline. This means that when you have to raise the rocker arm up to set the geometry, you are in essence moving the roller position farther across the valve stem tip.

Finally, longer rocker arms make no measurable difference in performance than a shorter rocker arm. Such as using the .150" longer big block Chevy rocker arm, on a small block engine. Exotic welding, and/or machining is a waste of time and money.

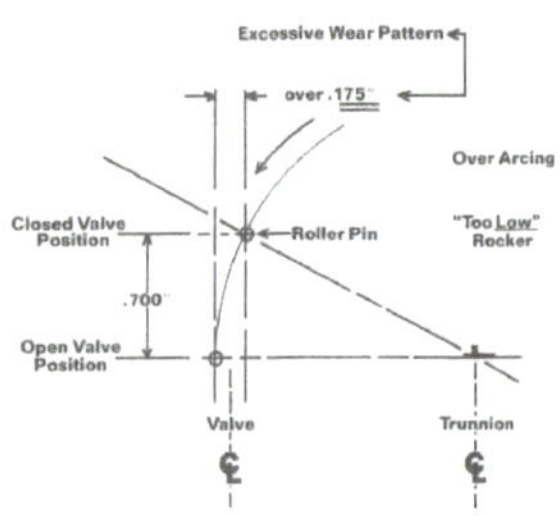

FIGURE 2-B

This is an individual representation of the "illustration A". (This Figure may be used in lieu of page A-1, Illus' "A")

HOW MID-LIFT™ EVOLVED

The principle of setting "***installed geometry***," originally all hand drawn in **1984**.

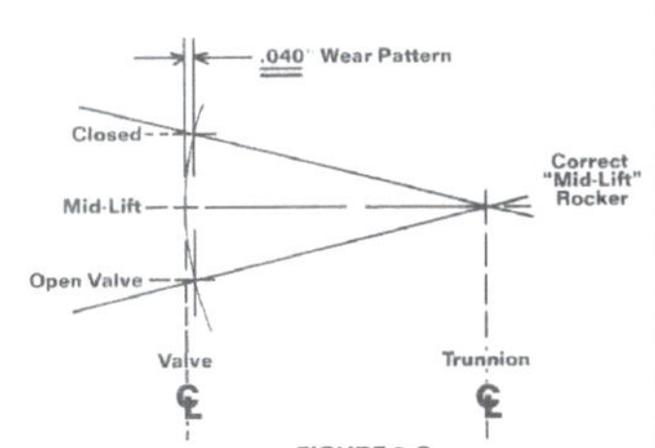

FIGURE 2-C

This is an individual representation of the "illustration B". (This Figure may be used in lieu of page A-1, Illus' "B")

In setting the geometry of the rocker arm on the engine, you must know what the NET valve lift is going to be. To do this, you will need to bolt the valve train together and turn the engine over, while accurately measuring this with a dial indicator. If you can manage to whip up an adjustable pushrod to handle this, it would be best. Once you know what your net lift is going to be, you can commence in locating the trunnion in the right place.

This "right" place, as mentioned, is the mid-lift point of the net valve lift. But this has to be in relation to the valve stem angle. This can easily be accomplished on the engine while assembled. The easiest and most accurate "aid" in doing this, is the valve spring retainer. Since it is exactly 90 degrees to our valve stem center line, it makes an excellent reference point to "aim" a machinist ruler along and past the center of the trunnion, while the rocker arm is in the closed valve (lash) position. Usually, there is a small nipple or indentation on the end of most trunnions to clearly mark its center. Place the ruler or a straight edge along the top of the retainer, and make note of the difference either above or below this center mark. Next, the height of the valve stem above the retainer top must be measured, either with a depth mike or calipers, and make note of it. Then finally, the diameter of the roller on your particular set of rocker arms, and divide by half. This final step will give you the "stack" measurement of the roller centerline on top of the stem height. This sum, when compensated by either adding or subtracting the trunnion location in relation to the retainer top in our first measurement, will give us the exact trunnion height with the roller centerline. Which in the closed valve position, should ideally be half of our NET valve lift.

For example: If the cam opens the valve .750", then our mid-lift point is .375". And if our straight edge passes .050" OVER the center of the trunnion. Then we would add our reference measurement of .050" to our stack measurement. Let's make the valve stem height .150" above the retainer top, and assume we are using a roller of .600" diameter. We simply add our half of the roller measurement of .300" to our stem height of .150", to our "plus" reference measurement of .050", to come with a total of .500". (If the reference measurement had placed the nipple .050" above the retainer horizon instead of below it, then the reference measurement would have been "negative," and subsequently subtracted from the "stack" measurement, rather than added.)

Since our mid-lift measurement is .375", we need to raise the rocker arm .125". Isn't that easy?

Naturally on engines with the conventional pushrods and guideplates, this height is set by the length of the pushrod. On shaft mounted assemblies, the shaft must be either spaced up with shims, or it must be milled down, as required. (Pedestals or stands)

This all applies to the regular "shoe" rockers as well. However, the measuring point is the contact surface of the shoe, in place of the roller pin centerline for the roller rocker.

If a change of cams is made, there may be a net valve lift difference of more than .040-.050", or the base circle of the new cam may be ground smaller, resulting in a lower sitting lifter. In either case, the pushrods should be substituted for the appropriate length. Also of course, if any of the other engine specs have been changed. (Deck height, etc.) Later. ●

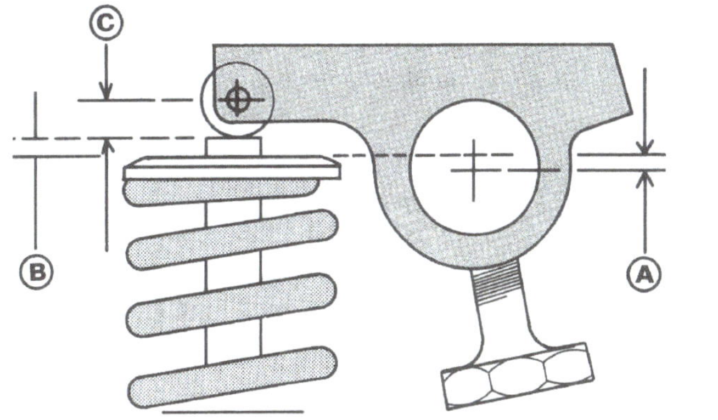

FIGURE 3-A

This is the typical aftermarket roller rocker arm as depicted in the "closed valve" position with its relative components of measurement. "A" represents the trunnion to reference line (straight across edge of the retainer top), while "B" represents the "stem tip" height above the same reference line. And "C" represents the roller pin centerline height (or one half of the roller diameter) which when added to "B", this sum becomes the "stack height" to which "A" is either added or subtracted (depending on whether it is below or above the reference line) for the final sum of the trunnion centerline below the roller centerline. As mentioned in the article, ideally this is equal to one half of the *total net valve lift*. Whatever the difference is, is the amount which the trunnion centerline needs to be changed. Don't forget however, that the figure the *trunnion* needs to be changed, *is not* proportionate to the length of the pushrod which will be doing the changing. The ratio of the rocker arm must be considered, but because the rocker arm is in a "closed valve" state during these measurements, the differing intersecting angles of the roller to trunnion vs. the trunnion to pushrod cup angle, reduces the full value of the ratio being applied to compute this. In other words, if you have a one and a half to one ratio small block chevy roller rocker arm, and your measurements show that the trunnion needs to be raised .100", then you cannot multiply the full value of the 1.5:1 ratio to your .100" requirement. It falls slightly less than this figure. What is the bottom line?.... Use an adjustable pushrod to raise as you measure the trunnion to its required "mid-lift" position. Then order your special length pushrods to fill the bill!

HOW MID-LIFT™ EVOLVED

Miller Engineering Inc℠

MID-LIFT®

Patented

Precision Roller Rocker Systems™

—THE Standard

By Which All Is Measured™

Professional Engine Builder

TECHNICAL EDITION

The following seven pages were printed in July of 1995, and disseminated to professional race teams, including NASCAR, under confidential agreements.
If you wonder where the NASCAR teams` leap to engineering their own valve-trains came from...
...wonder no more.

P950701
$5.00

Roller Rocker Arm Geometry

When it comes to roller rocker arms, one of the most misunderstood terms is "geometry," and it doesn't need to be.

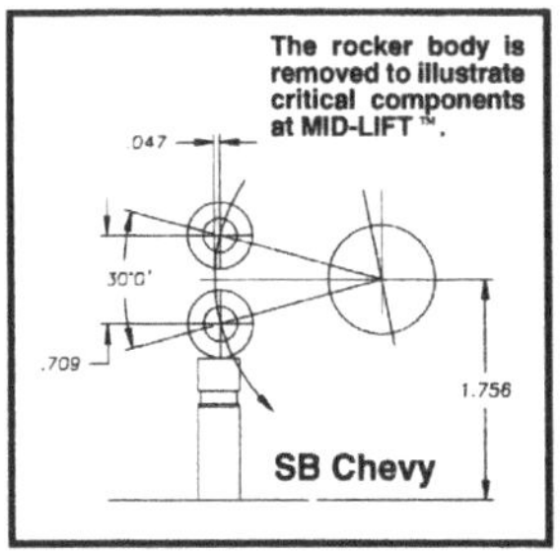

INSTALLED GEOMETRY

The illustration to the *left* is all that matters with the INSTALLED GEOMETRY. This pertains only with the valve side of the rocker arm, which can be adjusted by the engine builder through pushrod length (stud mounted) or stand height (for shaft mounted rocker arms).

The rocker's rollers, shown in the *open* and *closed* position with the shaft's pivot point and 30° arc.

The illustration to the *right* is representative of DESIGN GEOMETRY, which deals with the pushrod's relationship with the valve; correctly shown with radial effects on MILLER silhouette.

This simple, MILLER Patented concept and its importance has eluded the best companies of the valve train industry for 35 years!

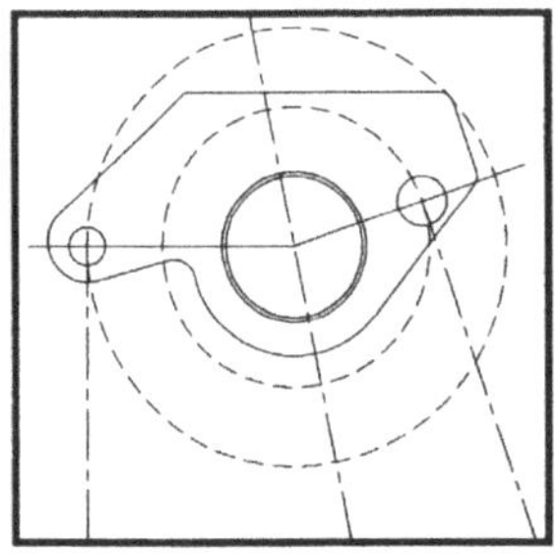

DESIGN GEOMETRY

What *Is* Rocker Arm Geometry?

Simply stated, rocker arm geometry is the *angles* of the two points of motion revolving around the rocker arm: the *pushrod* and the *valve*. It's NOT the height of the rocker arm, nor is it the length of the pushrods - these are merely consequences and methods to adjusting these angles.

What Angles?

There is an inherent angle to each engine which the pushrod leans into the valve. We call it the "attack angle". On some engines, the pushrod is inclined away from the valve, and here too the attack angle has critical influence and more limits to what can be changed.

Why Are These Angles Of Consequence?

Because these two angles are dictated by **LINEAR** motion being translated through **RADIAL** motion. In other words, they pivot around an axis in a circle to connect in-line components: the pushrod and the valve, respectively. If the intersection points of these angles are not matched to the specific angles of each engine's pushrod to valve design – then these intersection points and their related components will have EXCESSIVE motion on one end of the rocker or the other, or both. At the very least, they CHANGE the camshaft's instructions at the valve. If you move the height of the rocker arms up or down, you change the degree of time the valve and the pushrod see "in-line" motion. The roller spends *radial* time rolling across the valve stem tip, instead of *linear* time pushing down on the valve. Consequently, you change the degree of motion itself. In effect, changing both: the valve duration and the valve's lift.

The THIRD Dynamic

Not only is net valve lift and duration effected by improper geometry, but a third factor no one considers is also involved: the **RATE** *of acceleration*. When the height of a rocker's pivot point is changed, the rate of its acceleration, as given by the cam, changes at various points of lift. If the pivot point is high, the valve may accelerate quickly off the seat, then slow toward peak lift. If the pivot point is low, then the opposite can occur. It's quite possible to keep duration and lift the same, while completely juggling this third dynamic.

TWO Geometries

There are actually TWO geometries which must be separated when understanding how each applies to the engine and the rocker arm.

- (1) **DESIGN** geometry
- (2) **INSTALLED** geometry.

DESIGN geometry is the physical parameters the rocker arm is actually designed to, specifically: the *angles* by which both sides of the rocker arm's pivot points face one another, and intersect at the axis of the their pivot point: the trunnion/shaft. Center-to-center dimensions of the roller tip and the adjusting screw are secondary values. There is very little that any engine builder can do with the design geometry of a rocker arm; although adjusting screw rocker arms can actually change their design geometry by simply rotating the adjusting screw up and down. Too often, this is indiscriminately turned in to close the gap on short pushrods, and the consequences to this neglect can often cause excessive loading and damage from over-arcing and varied ratio. The choice of rocker arms must be careful, and must be done with insight to what you actually need.

INSTALLED geometry is the relationship of the rocker arm once installed on the engine, done by raising and lowering the rocker arm in relation to the valve's TIP. It seems to be an overlooked perspective because the rocker is sitting in a fixed position, but it is the valve's tip height – when the rocker is in the sitting CLOSED position which defines the limitations of how much the INSTALLED geometry can be adjusted. It's also a significant influence to our "design" parameters of rocker length - which - to a properly designed rocker arm, is a juggling act, especially considering the myriad of combinations in longer than stock valves, re-machined mounting surfaces, etc., etc. Here too, you must be aware of the effect you have on required rocker dimensions when you juggle these other components around. For instance, the rocker arm mounting stud or its mounting pad "leans" into the valve, and on the small block Chevrolet or Ford engines; for every .100" you lengthen the valve, you shorten this distance from the rocker's trunnion to the valve by approximately .020". When you go .200" or .300" longer than stock on your valves - you now have a considerably "too long" rocker arm.

The rocker arm is actually TWO rocker arms in one. Each side of the rocker arm is technically a "rocker arm". They must be designed collectively, but it's the pushrod tangents which dictate *everything.*

Design Geometry

PATENT 4,365,785
Additional Patents Pending

DESIGN GEOMETRY is the parameters by which the rocker body itself is designed. The Patented **MID-LIFT®** rocker arms duplicate the engine's pushrod to valve attack angles.

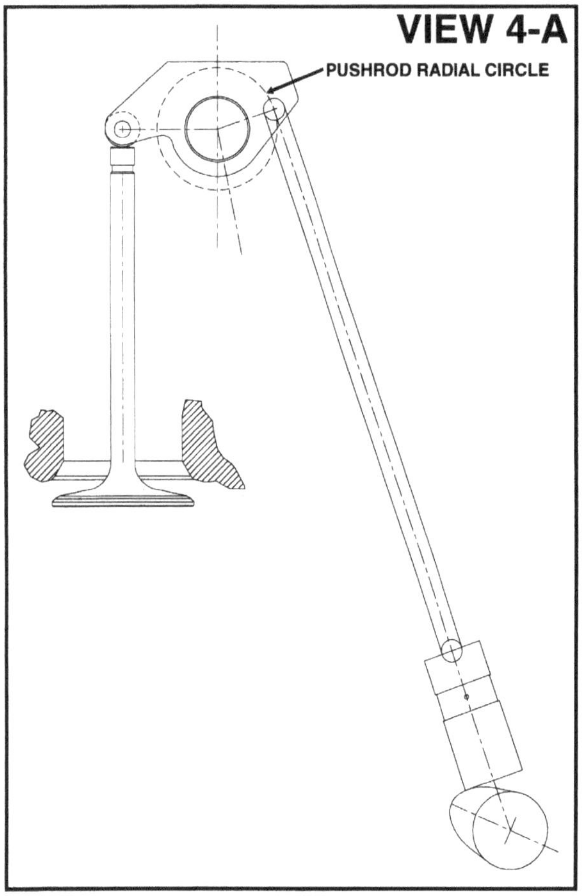

The **MID-LIFT "Design Geometry"** is determined by the pushrod attack angle, as measured with the valve.

VIEW 4-A: This CAD illustration, shows the MID-LIFT™ rocker arm and related components at mid-lift, or .314" (1/2 of the .628" valve lift). The large circle represents the actual pushrod cup's radial motion, which is the only critical lines shown in views 4-B and 4-C. It is *this side* of the rocker arm which determines all parameters of *design geometry* and its consequences.

The unique MID-LIFT® rocker body silhouette is shown here with the critical radial motion circles for both the *roller tip centerline* and the *pushrod cup centerline.*

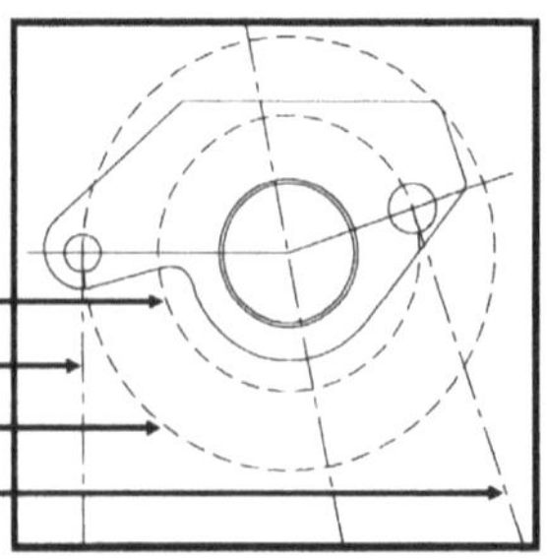

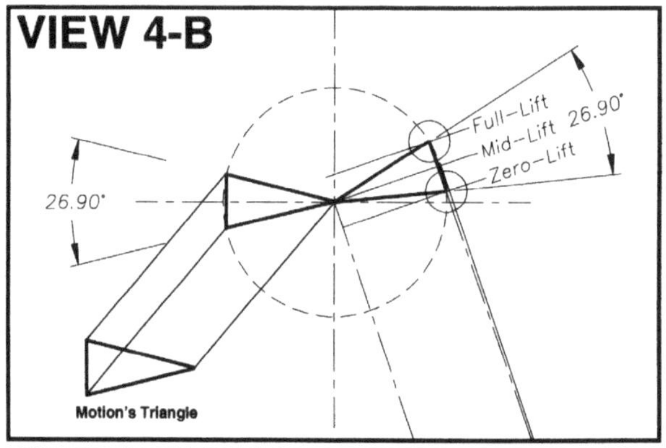

VIEW 4-B: Shows the radial circle with 3 reference lines intersecting the 3 stages of cam lift and rocker motion: Zero (closed valve), Mid-Lift and Full-Lift.

A *motion triangle* is formed (projected) which illustrates the rotation's critical start and stop points of the pushrod, as shown by the two small circles. This triangle is the rocker's *angular motion* —in this case: 26.9° is the result of the cam's *linear motion* of .400."

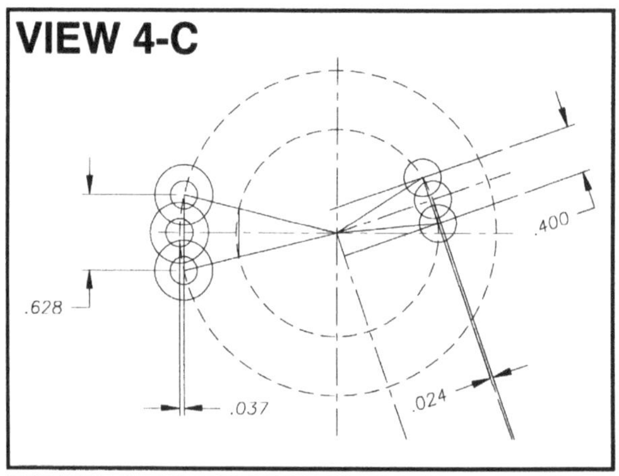

VIEW 4-C: Here, the effect of the 26.9° to its other side: the VALVE, represented by a *second*, larger radial circle, which is the length of the rocker arm's center-to-center between the roller and trunnion. (The roller and pushrod are shown in all 3 stages of valve lift.)

If the INSTALLED GEOMETRY sets the roller 90 degrees to the valve at the mid-lift point, then this same 26.9 degrees will translate from .400" cam lift to .628" valve lift. ***If the roller tip side starts at any other angle –the lift will be LESS.***

NOTE: *Radial sweep (back-and-forth) on this CAD illustration is only **.037"** on the valve, and **.024"** for the pushrod!*

Installed Geometry

INSTALLED GEOMETRY is the relationship of the rocker arm's pivot point (the trunnion or shaft) with the top of the valve stem, *as determined by the engine builder.*

VIEW 5-A

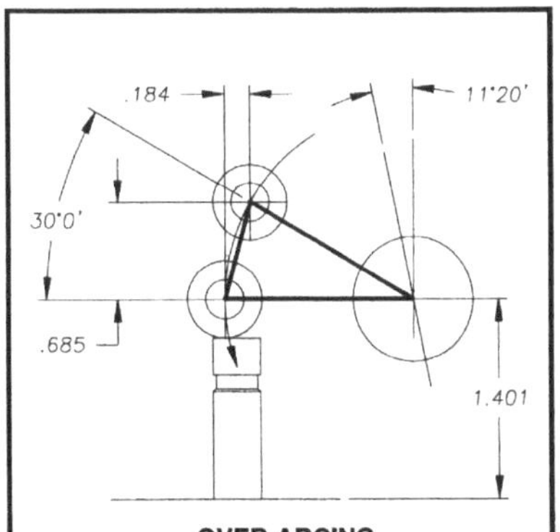

OVER-ARCING

VIEW 5-A: Here is the roller's sweep for a small block Chevrolet roller rocker arm, represented in an over-arcing (too low) position. The valve (tip only) is at the full lift position, which in the case is .685," as shown by the dimension intersecting the roller centers.

Note the angle of movement is 30 degrees. In this example the center-line of the trunnion and the lower roller is perpendicular to valve **at full lift**. This is an extreme over-arcing example, but it's often the case with many engines. The .184" dimension is the actual movement across the valve stem tip; while the 11° (degree) 20' (minute) figure is the actual stud incline into the valve for the small block Chevrolet.

As with the illustrations on the facing page, the **BOLD** lines form a triangle here: the valve side of the rocker arm, to show the angular range which the roller moves in going from its closed valve to peak valve lift. This "angle" was dictated by the *pushrod side*; but the NET valve lift is dictated by the ***angle*** of these *start and stop points*.

VIEW 5-B

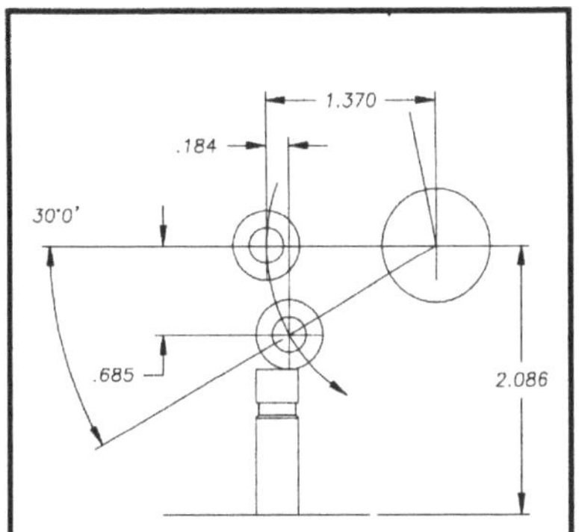

UNDER-ARCING

VIEW 5-B: In this view, we have exactly the opposite of VIEW 5-A. Here, the rocker arm starts out 90° to the valve center-line, then UNDER-ARCS across the valve, giving the same net valve lift, for the same 30° of motion. The rocker arm's roller tip to trunnion centerline is shown as 1.370."

The difference in height of the trunnion can be seen as the 2.086" figure, which is .685" higher than our example above. Keep in mind, that few people will actually move their rocker height around by this much, but for the purposes of example in cause and effect, these two extremes show how two opposites can have similar symptoms.

Over-arcing and under-arcing combinations which fall in between these two examples is easier to understand when you can see them exaggerated to these extremes. This valve side geometry (Installed Geometry) is totally dependent on the engine builder's choice of pushrod length and/or stand height.

VIEW 5-C: Now at **MID-LIFT**, the same 30° of motion now nets .709" of valve lift. An ***increase*** of **.024**" valve lift.

Here, **only .047"** *radial sweep* is used to move the valve **.709"**; compared to .184" for only .685" in the upper two views.

Although interesting, it is *NOT* the increase in valve lift that is important, as much as it is evidence of a symptom. The symptom is the LOSS of cam instruction through excessive radial motion across the valve stem tip. The same 30° of motion is taking place, but because the roller spends more time rolling *across* the valve in the upper two examples, the valve sees less "linear" motion; and also less DURATION.

As mentioned, this ANGLE of motion is actually determined by the pushrod side of the rocker; specifically by the angle of the pushrod's starting point in relationship to the trunnion axis.

Finally, take note of the two horizontal lines passing through the top of the STUD C-L, *representing .100" rise.* The decrease from 1.217" to 1.197" from the valve shows how raising the rocker or lengthening the valve reduces the distance between the stud and the valve. For every .100" increase in valve length, the rocker moves .020" further across the valve! This is why it's crazy to think that roller placement with the center of the valve is an accurate way to check for rocker arm geometry. ***The roller's contact point to the inside or the outside on the valve tip has no correlation to correct rocker arm geometry, and should NEVER be used for a guide.***

VIEW 5-C

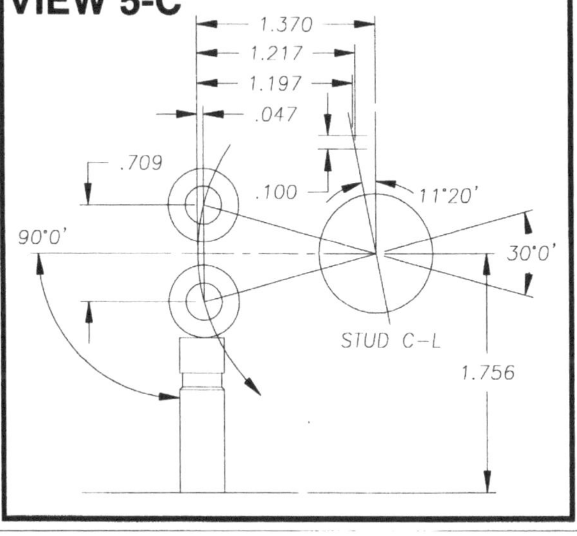

The Difference

MID-LIFT®

PATENT 4,365,785
Additional Patents Pending

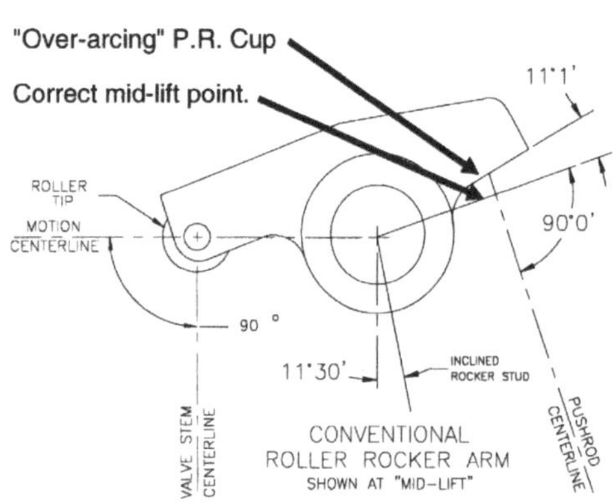

VIEW 6-1

Even though the above example of "*installed* geometry" has the "motion centerline" at mid-lift on the valve side of the rocker arm, this view shows the pushrod cup is much higher than the 90 degree line, where it *should be* for minimal in-and-out motion.

Because the "*design* geometry" of the above rocker arm is too high, the *pushrod cup* "over-arcs" – changing the cam's timing at the valve and increasing stud side loading and flex, regardless of setting the installed geometry to mid-lift with a longer pushrod.

Only the Patented MID-LIFT® rocker arms guarantee that the pushrod to valve angles of each engine are correctly duplicated by the rocker, holding pushrod motion to under .025" for an .800" valve lift cam!

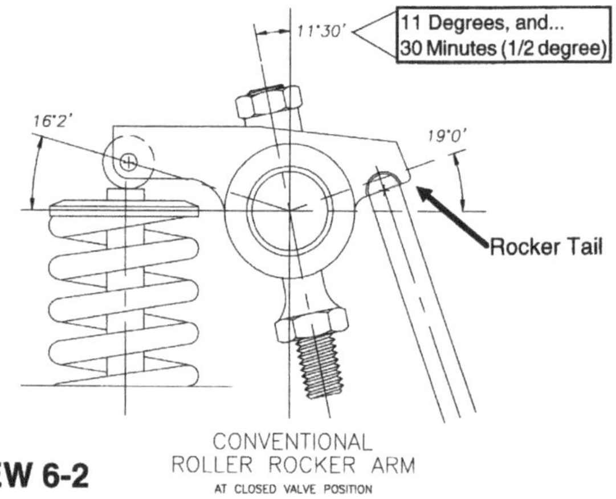

VIEW 6-2

This view, presenting the rocker arm in the closed valve position, shows one reason why the "typical" roller rocker arms were designed WRONG. Notice how the pushrod and the rocker tail are nearly 90° with each other. The pushrod is actually closer to the proper angle of operation when the valve is in the closed position.

The "originator" of the aluminum roller tip rocker arm approached his design in the "closed valve" position for all of the design criteria. His *1958 mistake* was copied by the top companies in the industry. Today, these leading companies continue to make the same mistake, with the similar incorrect relationship between the tail and the pushrod that is shown above.

The rocker arm's operating nature is "symmetrical." Except for different cam opening and closing ramps, the rocker arm can only *duplicate* in its closing the mirror reverse of opening the valve, regardless of its installed geometry. If the rocker has been set to accelerate quickly off the seat, then it will land quickly back on the valve seat, causing damage.

In addition to the rocker arm being a *radial* device which converts linear motion (pushrod) into radial motion, then back into linear motion (the valve), it multiplies this motion from one side to the other through its increased ratio. For this reason, to *design* or *install* the rocker arm in any way other than "splitting" this movement equally, will only increase the radial motion, and alter the valve's acceleration dynamics. Lift and duration can stay the same, but the valve's rate of acceleration off the seat can be entirely different, merely from changing the installed geometry. The valve "sees" a different cam!

MID-LIFT® geometry guarantees a balanced division of this motion, regardless of the valve lift. Anything else, will have the roller and/or the pushrod traveling out of their linear paths with excessive back-and-forth motion. When this occurs, the speed of the valve lift and where that speed accelerates through the lift cycle is changed, creating a different *cam* dynamic *at the valve*. And it's "**at the valve**" where the horsepower is realized; *not some arbitrary figures on a degree wheel at the crank, with cam figures whipped up on a computer.* —***jM***

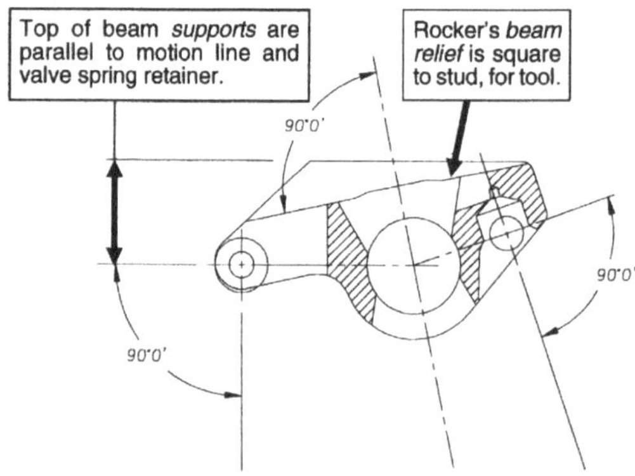

VIEW 6-3

The above view depicts the *correct* 90° balance of **both** sides of the patented MID-LIFT® rocker arm. Note the 90° relationship of the pushrod to cup.

The "**installed** geometry" is set to mid-lift by the engine builder, and since the "**design** geometry" of the MID-LIFT® rocker arm is correct, the PUSHROD is automatically 90° simultaneously. ***Everything is balanced.***

The MID-LIFT® rocker *body* is precision machined to match the angles of the rocker STUD and SPRING RETAINER, allowing the engine builder to set correct *installed geometry* easily and accurate with simple tools.

The Difference

MID-LIFT®

When MID-LIFT® geometry has been achieved, the radial travel across the valve stem tip will be less than .040" for .650" valve lift, and less than .060" for up to .850."

On page 5, two illustrations show how the roller can travel across the valve stem by as much as 1/2 of the stem diameter, or more.

CLOSED VALVE

Roller contact is to the inside of the valve's centerline.

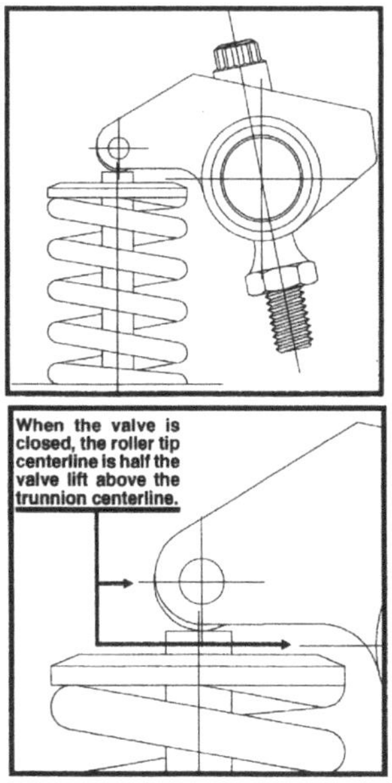

MID-LFT®

Roller contact is to outside of the valve's centerline, at its furthest.

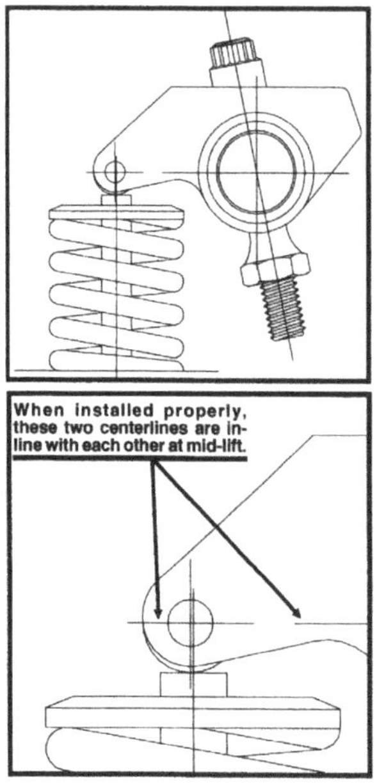

FULL-LIFT

Roller contact is to the same inside point of the valve's centerline as in the closed position.

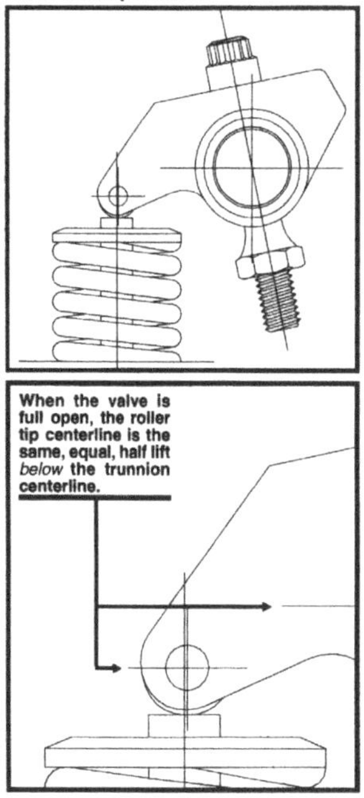

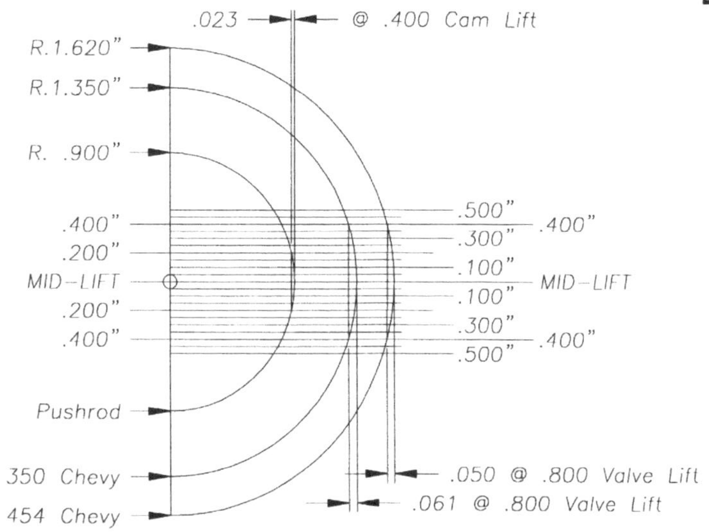

In the graph above, the in-and-out motion of the Small Block 350 and Big Block 454 Chevy rocker arm is represented at .800" valve lift, to the right side of the graph, while the *radial movement* of the pushrod is shown on the left side of the graph.

NOTE the in-and-out motion of the pushrod is only .023" for its lifter motion of .400". The radial in-and-out motion shown is the ultimate minimums for these rocker arm dimensions, which are ONLY attainable at a 90° measurement to the valve.

Stud Versus Shaft Choices...

Over the last decade, many engine builders have chosen to use shaft mounted systems for engines, which originally were designed for **STUD** mounted rocker arms. In some cases, this has been an improvement; but in some cases it has not. Often, their decision has been a result of "advertising" —more than need.

There are merits to both designs, which is why we design shaft systems for most stud mounted applications, in addition to the stud designs which we offer. Regardless which version we make, the main theme of design comes down to one basic factor: ***SIMPLICITY***. The most simple approach is always best. Our end product may not look so simple, but the principles surrounding its design are. We approach each design for *minimum motion with the least amount of energy loss*. When all factors of each application is fully considered (attack angles, side thrusts, unit loads), achieving this criteria will always produce the best product.

The following two pages try to give a little credit to STUD systems, or at least – put the stud designs in perspective. *If you want the ultimate rocker system that we make, SHAFT is going to be the way to go... but it's NOT ALWAYS necessary;* nor do we push it when it's not. As long as the geometry of the rocker is correct – stud often works fine.

With rocker arm and pushrod valve train there are TWO variables of *valve layout* which influence the ultimate system. These are two distinct designs which follow either a **2-PLANE** or **3-PLANE** array.

- 1. **2-PLANE** systems have their valves in-line along one axis in relation to the cam axis; either a straight line from an end view of the head, like the Small Block Chevrolets, Chryslers and 302 Fords (the X-axis), or the less common design of the Chrysler Hemi which has the valves in-line, but perpendicular to the camshaft (the Y-axis) with its rockers mounted on an in-line shaft.
- 2. **3-PLANE** systems, are "canted" valve designs that combine a third angle which falls in between X and Y. The Big Block Chevrolet, the Ford Cleveland, 460 wedge and the opposing valve design of the BOSS 429 Ford are good examples. All 3-PLANE systems require "compound angles" of the *pushrod* to the *lifter* centerline in providing port layout. It is the pushrod and lifter alignment which are the first denominator in deciding the "simplicity" in which a rocker system design can be approached.

The canted valve and hemi engines, due to their port layout and compound angle pushrods, all have unique limits on where and how the rocker arms are mounted, making a valve train design that is capable of going both ways difficult and counter-productive. *With these original stud mounted engines, any shaft design which merely bolts a single stand and shaft to the original stud mounting location is a compromise which really accomplishes little.—jM*

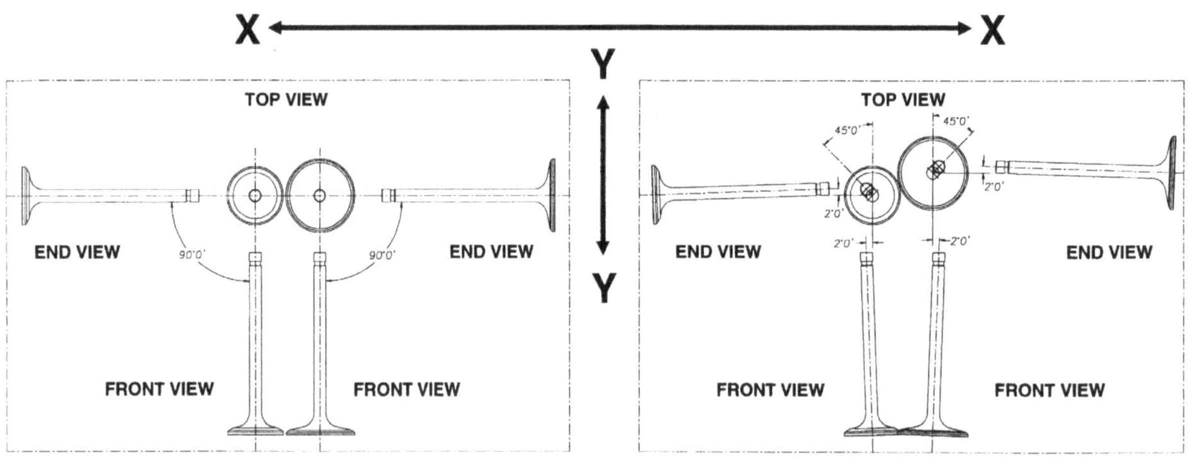

VIEW 8-A

2-PLANE:
VIEW 8-A is the most simple and proven performer, due mainly to its ease of valve train compatibility. Everything can be designed to parallel the camshaft, without the wasted motion from compound angle pushrods and rocker arms.

The "X" and "Y" arrows shown, represent only the axis of the TOP views of the above drawings. The "X" axis is the same as the CAM. The "Z" axis is the valve's up and down motion.

VIEW 8-B

3-PLANE:
VIEW 8-B is typical in principle of the Big Block Chevrolet and Cleveland Ford, to name only two, although their actual angles differ. In our example the valves have been shifted 2° on both X and Y planes, which create a THIRD PLANE that is at a 45° angle. *Depending on the pushrod angle leaving the lifter*, the rocker arm must be mounted at a complex and accurate angle to maintain full contact on the valve stem tip, without deviating from the valve's center-line during valve lift. Stud designs simplify this chore.

Shaft & Stud
Hype & Myth

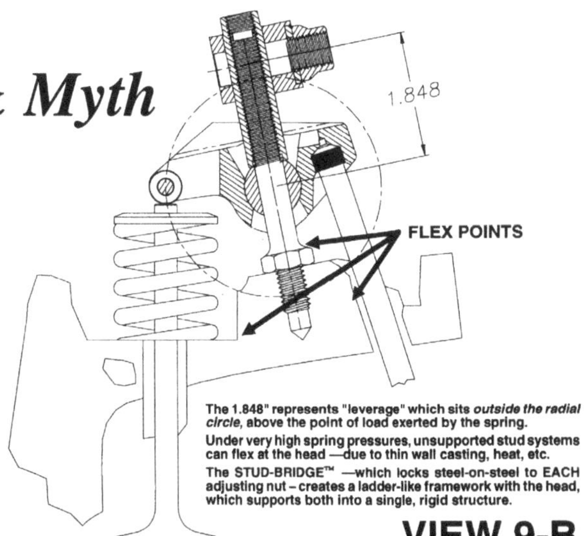

The 1.848" represents "leverage" which sits *outside the radial circle*, above the point of load exerted by the spring.

Under very high spring pressures, unsupported stud systems can flex at the head —due to thin wall casting, heat, etc.

The STUD-BRIDGE™ —which locks steel-on-steel to EACH adjusting nut – creates a ladder-like framework with the head, which supports both into a single, rigid structure.

VIEW 9-B

Shaft systems *are better than stud designs because they:*

- **1**. Remove rocker weight from valve, by resting on shaft.
- **2**. Reduce flexing of the stud, by removing the stud mounting.
- **3**. Increase horsepower with longer, offset rockers.

WRONG, WRONG, WRONG.

With regard to #1: see "Stud Mounted Rocker Weight".

With regard to #2: too much blame is given to the flexing of studs, when flexing also occurs at the head's stud boss. If a shaft system is going to substitute the original stud system, and does not take advantage of tieing these mounting points together, then the next best thing is to retain the leverage and hold the rocker assembly by its stud with a high *handle point* ABOVE the valve spring, and in-line with its loads. (See: VIEW 9-B). Correct stud design rockers, with good stud girdles take advantage of resisting flex from a higher leverage point above the spring forces, and work.

It makes little sense to take a cylinder head that has been specifically designed with a single, stud mounted rocker arm, and then mount a "stand" to it - with a single bolt - that then uses two bolts to hold a "shaft" - which then holds the rocker arm!..? All this additional material, still, ultimately attached to the head with ***only one bolt*** *screwed in where the stud was in the first place. The only argument to be had is that the "rocker arm weight" has been removed, because it certainly can't be "flex." Flex is supposedly controlled by then securing the stands together with another bolt, welding, or not at all. Unfortunately, this solves little.* ***There is a better way!***

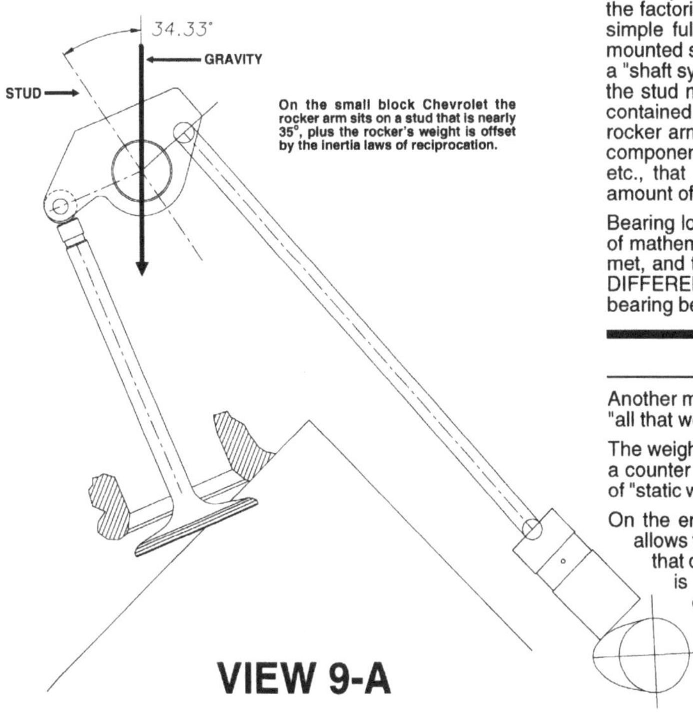

On the small block Chevrolet the rocker arm sits on a stud that is nearly 35°, plus the rocker's weight is offset by the inertia laws of reciprocation.

VIEW 9-A

With regard to #3: longer rocker arms were substituted to acquire enough room under their roller beam to allow notching for larger valve springs without excessively weakening the rocker body, which is justified. But we learned long ago that reduced drag through longer rockers was really the result of changed operating geometry. The real factors which are influenced are beam loads on the rocker body itself; and only in extreme examples does increased lengths and reduced "angular acceleration" offer any improvements alone.

- Whether you call the bearing device your rocker arm runs upon, a *trunnion, fulcrum or shaft*, what really matters is: geometry, motion and load. The **simplest** way to secure these three criteria, is ***the best*** way to go.

Shaft system rocker arms have taken on several meanings since the factories early designs, decades ago. Today, these can entail simple full-length assemblies across all cylinders, to individually mounted shafts on each rocker arm. All ROLLER rocker arms are a "shaft system" if you get truly technical about it. The "trunnion" of the stud mounted systems *is* a "shaft" – even if it's merely a self contained, single unit. The considerations of a properly designed rocker arm system, to a designer, have to do with unit load to the components, and the required bearing sizes and bolt, studs, stands, etc., that is all necessary to control these forces with the least amount of FLEX and WEIGHT.

Bearing loads within the rocker arm itself are simple, and a matter of mathematics. As long as the area of loading required has been met, and the integrity of the rocker body maintained, it makes NO DIFFERENCE if there is one stud between two bearings, or one bearing between two studs.

Stud Mounted Rocker Weight

Another misleading issue is the stud mounted rocker arm imposing "all that weight" on the valve.

The weight is RECIPROCATING (back and forth), and as such has a counter inertia effect which cancels itself to the simple argument of "static weight," allowing very little to actually be seen by the valve.

On the engine, these components sit at an angle, which in part, allows the trunnion and rocker arm to *lay* on the stud at an incline that diminishes straight down free weight. The little weight that is actually seen by the valve, is not to the point of being detrimental to the cam or valve train's efficiency, except *maybe* in the most extreme of conditions. 9,000 rpm engines have been setting records for 3 decades on *mere* stud mounted rocker arms *—and they will continue to do so. —jM*

About The Author

Jim Miller has been involved in the racing industry for more than 40 years. Like many company founders in America's racing heritage, Jim's roots stem from amateur racing at 16, after having begun "building engines" at 13. By 17, Jim was the youngest Ford trained line mechanic authorized to do warranty work on all of Ford's factory muscle cars. He'd already spent two years building and racing an 11-second 1967 Mustang 390 GT, that was still driven to High School during the week, while *Steve McQueen* made the exact same 1968 version famous in *Bullitt*.

Before Jim's 18th birthday the Mustang GT would be traded for a brand new 1970 BOSS 429 Mustang. Within a month it was disassembled and prepared for NHRA's B/MP using the best parts from Holman-Moody, Gapp & Roush, and other top names. Jim quickly acquired a reputation for high performance ingenuity that belied his youth. By 20, he'd already developed MID-LIFT rocker arm geometry, and *Dyno Don Nicholson* had asked Jim to move to Georgia to be Chief Mechanic for his *Pro Stock Maverick*. Having lost his father that year, Jim respectfully declined, and as a result the opportunity was passed on to Jon Kaase, who built a career around big inch Ford engines.

In **1973**, Jim began developing his first rocker arms to alleviate problems with arguably the most complex compound geometry rocker system of any American OHV engine, the BOSS 429. It required a complete study of rocker systems on America's most popular racing engines, and led to questions no one else seemed to ask. Those answers became the principles of MID-LIFT geometry.

In **1975**, as one of SuperFlow's earliest customers, Jim's focus on cylinder head development expanded to include work for other competitors from all over the Midwest.

Miller Race Engineering (**MRE**) was soon opened where Jim received national notoriety when his byline appeared in a variety of top magazines publishing his in-depth technical articles on cylinder head development, engine preparation, and of course valve-trains. These included *Popular Hot Rodding, Super/Stock, Pulling Power,* and *Super Ford,* among others. An avid Harley-Davidson rider and custom show bike builder, Jim also penned for various bike publications, including *Custom Bike* and *Choppers*.

In **1980**, with a Patent Pending, and 5 years experience on his own engines after 2 generations of rockers that included *Bob Glidden* among his customers, Jim published a full page ad in *National Dragster* illustrating the principles of MID-LIFT™ geometry and attracting the attention of many companies, including TRW's valve division.

In **1982 the MID-LIFT™ Patent (#4,365,785)** was issued, spelling out perpendicular geometry at mid-lift used the very least amount of wasted motion in converting linear motion from the cam through the rocker arm to the valve.

In **1983**, Jim had refined rocker arm "beam" design and side thrust loadings to what became the 3rd Series of BOSS 429 rocker arms. Rocker arm "firsts" that were exclusively Jim's innovation was needle thrust bearings on the side loaded rocker bodies; precision length adjusting screws and matching height nuts for precisely setting cup position, often referred to as "two threads out," and rocker body measuring faces that matched stand tops at MID-LIFT to allow engine builders an easy way to set *installed geometry*. Customers included some of the biggest names in racing, including World Champion Bob Glidden and Funny Car World record holders, the Jordon Brothers.

In **1984**, Jim published a detailed story on *Installed Geometry* in *PULLING POWER Magazine*, which laid out his hand drawn rocker arm, valve spring and retainer illustration that would become a standard of use for thirty years.

About The Author

By **1988**, Jim joined Lunati Cams in Memphis to oversee new product development, implement a MID-LIFT product line, and redesign an all new Lunati catalog where Lunati's peers in the valve-train industry were formally introduced to the MID-LIFT *installed geometry* illustration, which Jim now done in CAD. The new catalog format which placed more emphasis on the cam specifications than any other company was acclaimed as: *designed by an engine builder for engine builders*. Crane Cams interceded in the Miller/Lunati rocker merger with "an offer Lunati couldn't resist," and dedicated MID-LIFT rocker arms would not be made. As a result, Jim left Lunati Cams and the racing industry in 1989 to pursue other passions that took him to Miami in 1992. But Jim's *installed Geometry* CAD illustrations, instructions, and the new catalog format Jim designed and wrote remained for over a decade.

July, 1993, Jim returned to the racing industry and founded *Miller Engineering Inc*, solely to make a complete line of MID-LIFT™ geometry rocker systems. To introduce the technology in detail, in July of 1995 Jim wrote the confidential *Professional Engine Builder Technical Edition* brochure for race teams under confidential agreements, and later that year a second company, *Precision Valve Systems* (PVS) was formed to distribute a more affordable, MID-LIFT geometry rocker line for warehouses where engine builders could make a reasonable profit. For PVS alone, Miller Engineering developed numerous pioneering product lines, including a stamped chromoly steel shoe tip rocker and roller tip version; a 7075 aluminum stud mount rocker with the first ever gun-drilled trunnion, a Patent pending G-TooL™, and the Stud-Bridge™. The high-end professional stand mount NASCAR and Billet Stud rockers made by Miller Engineering were also unprecedented, patented or patent pending, and made in-house on the finest Kitamura and Mori-Seiki multi-axis pallet changing CNC mills and lathes money could buy. Miller Machine, the manufacturing arm of Miller Engineering, founded in 1998, would go on to do job-shop manufacturing for diverse medical companies as well as the aerospace giant, Pratt-Whitney.

From 1995 thru 2001, Jim confidentially consulted for several of NASCAR's top race teams developing MID-LIFT rocker systems and related cam alterations to maximize the precision geometry. During this time Jim was awarded three (3) more Patents: ***#5,560,265; 5,596,958 & 6,041,750***; with others pending, which Jim now expands in 2016.

In 2003, the **MID-LIFT.com** web site was launched, where Jim illustrated rocker geometry and cylinder head integration details never before published, to support his PVS distributors and engine builders receiving the conflict of information that established, big name cam and rocker arm manufacturers were using to still denounce Mid-Lift as a "theory." During this time several of PVS's warehouse distributors began undercutting the price structure Jim implemented for engine builders. So on the heels of a recent and expensive divorce that distracted ten years of work building the PVS product line, it was dismantled, with many high demand products discontinued in 2004 and '05.

In 2007, following three years of costly growth pains to ramp up production, banks began to pull back ahead of the financial industry collapse, and MEI was forced to close after a costly breach of contract by the building owner.

A year later, Jim retired from manufacturing and returned to his long standing passions with animal conservation, aviation, film and feature writing projects unrelated to the racing industry; while continuing to support MID-LIFT.com.

In 2010, the *Advanced Tech Bundle* was published; a compilation of detailed rocker arm tech available by PDF download from the MID-LIFT.com web site, which ***MID-LIFT™ Advanced Rocker Arm Tech*** replaces for **2016**, marking Jim Miller's final automotive byline.

EPILOGUE

In 1998, **Crower** quietly changed their decades old steel rocker (*left*) to be closer to Mid-Lift principles (*far right*), without even changing the part number; then, they copied the MID-LIFT™ drilled trunnion.

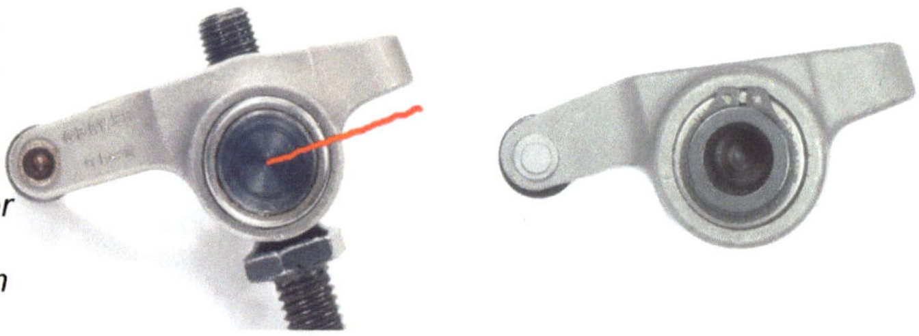

When Dave Crower was confronted in **1998** about infringing on the **MID-LIFT**™ Patent, *Bruce Crower* denounced mid-lift in a letter as his company was copying it; then, hijacked "Half Lift" as *their version* of "***Engineering the Science of Geometry***."

...Glad I could help!

CROWER
Cams & Equipment Company, Inc.
3333 Main Street, Chula Vista, CA 91911-5899 USA
Phone: 619-422-1191 • Fax: 619-422-9067

Miller Engineering Inc.
1791 Blount Rd, 501
Pompano Beach, FL 33069

March 2, 1999

Re: United States Patent 4,365,785

Dear Mr. Miller,

Loren Harris asked me to contact you with regard to your letters to him and Dave Crower.

You have an opinion as how rocker arms should be made, and have a patent on that opinion. In the opinion of both myself and Loren Harris, the position of the pushrod upper ball axis should not be in the mid-lift position, or even in the range you claim still represents perpendicularity.

Also, on the valve side of this lever, we do not use the axis of the roller for design purposes, as it has no function but to carry the moving surface that contacts the valve stem.

I have no interest in debating these matters by letter, telephone, or in the court room. However, if you should choose to use a court of law in this matter, I am prepared to defend our position.

Sincerely,

Bruce Crower

You have an opinion as how rocker arms should be made, and have a patent on that opinion. In the opinion of both myself and Loren Harris, the position of the pushrod upper ball axis should not be in the mid-lift position, or even in the range you claim still represents perpendicularity.

Also, on the valve side of this lever, we do not use the axis of the roller for design purposes, as it has no function but to carry the moving surface that contacts the valve stem.

www.ingramcontent.com/pod-product-compliance
Lightning Source LLC
LaVergne TN
LVHW070951210926
848482LV00002B/185